ENVIRONMENTAL CHEMISTRY

First Published – 1994

Reprinted – 2017

ISBN: 978-81-7141-242-6

Environmental Chemistry

Published by:

DISCOVERY PUBLISHING HOUSE PVT. LTD.

4383/4B, Ansari Road Darya Ganj
New Delhi - 110 002 (India)
Phone: +91-11-23279245, 43596064-65
Fax: +91-11-23253475
E-mail: discoverypublishinghouse@gmail.com
sales@discoverypublishinggroup.com
web: www.discoverypublishinggroup.com

Printed at:
Infinity Imaging Systems
Delhi

ENVIRONMENTAL CHEMISTRY

M. SATAKE
Faculty of Engineering
Fukui University
Fukui 910, Japan

Y. MIDO
Dept. of Chemistry
Kobe University
Kobe 657, Japan

Consulting Editors

S.A. IQBAL
Dept. of Chemistry
Saifia P.G. College of
Science and Education
Bhopal, INDIA

M.S. SETHI
A.R.S.D. College
University of Delhi
New Delhi-INDIA

Discovery Publishing House
New Delhi (INDIA)

Preface

The teaching of Chemistry at the introductory stage becomes each day a more challenging task as the subject matter becomes more diverse and more complex. These challenges have evoked a series of responses – the present set of introductory chemistry monographs is one such. The teaching of chemistry recognises a number of problems that confront those who select text books. In order to overcome these problems, this volume "**Environmental Chemistry**" – one of about fifty in the Chemistry Monograph Series – is introduced. Each volume is independent of the others deals with one of Chemistry topics and constitutes a complete entity. Each volume is more comprehensive than can be possible in a single volume text. It is intended to provide a range of topics to cover most undergraduate and chemistry main courses of study. These volumes can be used to enrich the more conventional courses of study.

Suggestions for improvement are welcome and shall be gratefully acknowledged.

Authors

Contents

CHAPTER 1

Introduction

THE ENVIRONMENTAL SCIENCES

Environmental issues have become a matter of widespread public concern only over the past twenty year or so. Nonetheless, basic environmental science has existed as a facet of human scientific endeavour since the earliest days of scientific investigation. In the physical sciences, disciplines such as geology, geophysics, meteorology, oceanography, and hydrology, and in the life sciences, ecology, have a long and proud scientific tradition. These fundamental environmental sciences underpin our understanding of the natural world and its current-day counterpart perturbed by human activity in which we all live.

These are two major reasons why environmental chemistry has come up as a discipline only rather recently. Firstly, it was not previously perceived as important.

The idea that using an aerosol spray in your home might damage the stratosphere, although obvious to us today, would stretch the credibility of someone unaccustomed to the concept. Secondly, the rate of advance has in many instances been limited by the available technology. Thus, for example, it was only in the 1960s that sensitive reliable instrumentation became widely available for measurement of trace concentrations of metals in the environment. This led to a massive expansion in research in this field and a substantial *downward* revision of agreed typical concentration levels due to improved methodology in analysis.

THE CHEMICALS OF INTEREST

The chemical substances considered in this book, fall into three main categories:

(i) Chemicals of concern because of their human toxicity. Thus, for example, the metals, lead and mercury are well known for their adverse effects at high levels of exposure. Chemical carcinogens are particularly topical, despite the miniscule risks associated with many of them at typical levels of exposure. Examples are benzene (largely from vehicle emissions) and polynuclear aromatic hydrocarbons (generated by combustion of fossil fuels)

(ii) Chemicals which cause damage to non-human biota, but are not believed to harm humans at current levels of exposure. PCBs are an example, believed to be having a serious effect on some animal populations, but not proven to be harmful to humans.

(iii) Chemicals not directly toxic to humans or other biota, but capable of causing environmental damage, *e.g.* CFCs.

THE ENVIRONMENT AS A WHOLE

Traditional boundaries between atmosphere and waters are not a deterrent to the transfer of chemicals (in either direction), and many important and interesting processes occur at these phase boundaries.

In this book, fundamental aspects of the science of the atmosphere, waters, and soils are described together with current environmental questions. Subsequently, quantitative aspects of transfer across phase boundaries are described. Monitoring considerations, the effects of chemical pollution are considered next.

CHAPTER 2

The Atmosphere

THE STRUCTURE OF ATMOSPHERE

Troposphere and Stratosphere

The vertical structure of the atmosphere, is illustrated in Fig.2.1. The depth of the troposphere is 8-15 km, the lowest values occurring at the poles and the highest at the equator. There is also a variation with season of the year. Within this layer occurs most of the variability of conditions which leads to 'the weather' as the layman experiences it. Above the troposphere lies the stratosphere which is relatively cloud-free and considerably less turbulent—hence long distance passenger jets fly at altitudes corresponding to the top of the troposphere. The distinction between these two layers is based on a change in the temperature variation with height (Fig.2.1). Within the troposphere temperature decreases with height but as we enter the stratosphere the temperature starts to increase again. This point is termed the tropopause. The situation with a layer of warmer, less dense air over a layer of cooler, denser air is quite stable. Consequently air is exchanged between the troposphere and stratosphere only very slowly.

'Air pollution' normally means pollution of the troposphere within which most pollutants have a fairly limited lifetime before they are washed out by rain, removed by reaction, or deposited to the ground. However, if pollutants are injected directly into the stratosphere they can remain there for long periods resulting in noticeable effects over the entire globe. Thus major volcanic eruptions injecting fine dust into the stratosphere can lead to a reduction in the amount of solar energy reaching the ground for more than a year after the event. This could also be the result of a major nuclear conflict and has led to fears of a 'nuclear winter' which could make life extremely difficult for survivors.

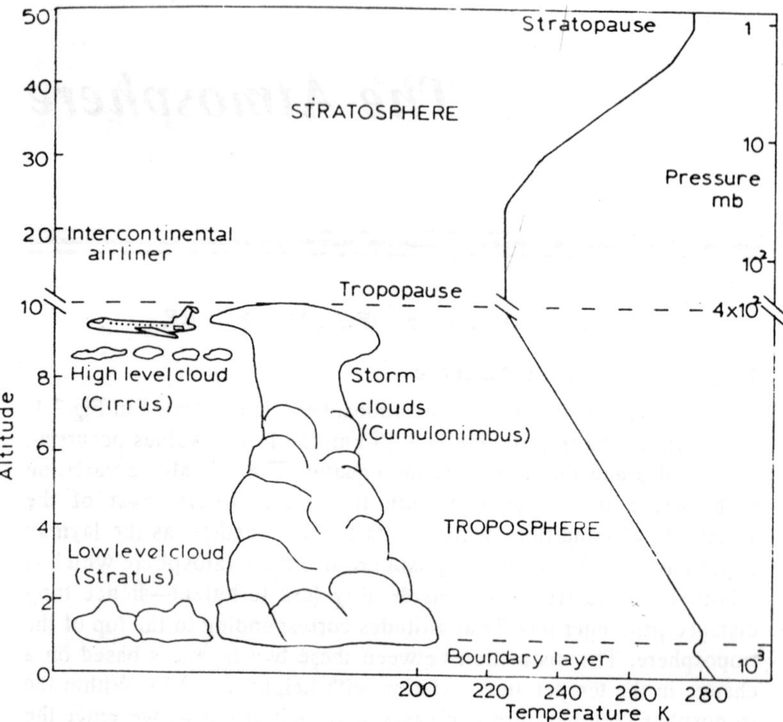

Fig.2.1 *The vertical structure of the atmosphere. The temperature profile would be typical for latitude 60°N in summer. Note the change of scale used for the upper half of the figure*

Atmospheric Circulation

The main driving force for the circulation of the atmosphere is the incident solar radiation. The solar energy falling on a given area varies with latitude so that the poles are cold and the equatorial regions warm. The proportions of the incident energy reflected back to space, absorbed by the land or sea, and re-radiated at a longer wavelength all vary from place to place and affect the temperature distribution and circulation patterns. This energy balance is crucial to the determination of the global climate and is considered in more detail in the next section. The rotation of the earth affects the circulation patterns in a fundamental way resulting in the tendency of air to circulate in large-scale eddies around the 'low ' and 'high' pressure regions on synoptic weather charts. The processes of evaporation of water, cloud formation, and precipitation also affect the energy balance and circulation patterns.

The presence of the ground has only a small effect on the overall pattern of atmospheric circulation and at most altitudes air movements approximate to those of a non-viscous fluid. The theoretical wind speed—the so-called geostrophic wind speed—can be calculated from the pressure gradient and the rotational velocity of the earth. The pressure gradient is reflected on a weather chart by the closeness of the isobars, lines of constant pressure. If the isobars are close together the wind speed will be high.

The Boundary Layer

Near to the ground the situation is more complicated. Turbulence is generated by mechanical forces as the air flows over uneven ground features such as hills, buildings, or trees and also by buoyancy forces. The ground may warm or cool the air next to it resulting in up-currents and down-currents. There is a frictional effect of the ground. Consider the variation of wind speed with height over the lowest few hundred metres of the atmosphere. This variation is greatest over rough surfaces (*e.g.* a city) when the effect could be a reduction of 40% of the wind speed aloft, that is, the geostrophic wind. Over smooth surfaces (*e.g.* sea, ice sheets) the effect is less and the reduction may be only 20%. Frictional effects also result in a variation of wind direction with height—'wind shear'. A plume from a tall chimney may appear to be travelling at an angle to the ground level wind.

Within the troposphere we therefore define a boundary layer within which surface effects are important. This is of the order of 1 km in depth but varies significantly with meteorological conditions (Fig.2.1). Vertical mixing of pollutants within the boundary layer is largely determined by the atmospheric stability which relates to the intensity of buoyancy effects previously mentioned. As a generalization, mixing within the boundary layer is relatively rapid whereas mixing through the remainder of the troposphere is slower. This gives rise to the idea of a mixing depth within which pollutants are retained and may be transported long distances.

Table 2.1 indicates the time and distance scales involved in the dispersion of pollutants emitted from the ground. No account is taken in this table of the rates or removal of any pollutant by reaction, deposition to the ground, *etc.*

Table 2.1 Time and distance scales for atmospheric dispersion of emissions

Time of travel	Typical distance	Area affected
Hours	Tens of km	Throughout boundary layer.
Days	Thousands of km	Pollutant escaping from boundary layer into free troposphere.
Weeks	Round the earth	The whole troposphere in one hemisphere. Transport to other hemisphere beginning.
Months	Round the earth	Whole global troposphere. Some penetration into lower stratosphere.

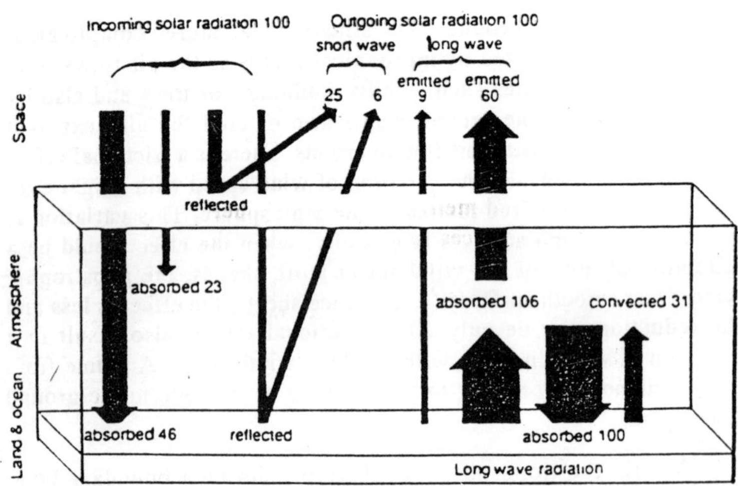

Fig.2.2 *The energy balance of the earth expressed in terms of percentages of the total incoming or outgoing solar radiation.*

CARBON DIOXIDE AND THE GLOBAL CLIMATE

The Global Energy Balance

If the earth had no atmosphere, the mean surface temperature would be 255 K, well below the freezing point of water. The atmosphere serves to retain heat near the surface and the earth is thereby made habitable. Most of the radiant energy from the sun lies in or near the visible region of the spectrum (*i.e.* at short wavelength *ca.* 0.6 μ m) . Some light is reflected unchanged from clouds or from the ground (especially by snow or ice). The fraction of reflected light is termed the albedo and is over 0.5 for clouds but below 0.1 for the oceans. The global average albedo is about 0.3. The stratosphere absorbs ultra-violet radiation primarily due to the ozone present and this results in the warming shown in Fig.2.1. The lower atmosphere is transparent to visible light so it gains relatively little energy from incoming radiation. The transmitted radiant energy in the visible region penetrates to the ground and is absorbed. Fig. 2.2 shows the percentages of the radiation for different components of the overall energy balance. The radiation emitted from the ground lies in the infrared region of the spectrum (long wavelength, *ca.* 10-15 μ m) and several atmospheric constituents absorb in this region. Carbon dioxide, water vapour, and ozone are the most important of these. Methane, nitrous oxide, and chlorofluorocarbons (CFCs) are also significant. Some of the absorbed energy will still be re-radiated back to space but a part will be returned to the ground or retained in the atmosphere. The final factor that results in surface to atmosphere transfer of energy is direct warming of the air nearest the ground together with evaporation/condensation processes. The net effect is that more energy is retained near the surface of the earth and the mean temperature is therefore higher (global average 288 K). This is sometimes described as the 'greenhouse effect' by analogy with the properties of glass. Glass is largely transparent to solar radiation while absorbing completely radiation in the infra-red at wavelengths greater than 3 μ m. In fact the most important function of a greenhouse is to prevent the circulation of air, inhibiting the normal cooling processes, but the term 'greenhouse effect' has nonetheless been retained.

The Carbon Dioxide Cycle

There is no doubt than man's activities are leading to a gradual increase in the atmospheric carbon dioxide level and this leads to the prospect that we may eventually modify the global climate. Fossil fuel burning is the main contributor to the global annual emissions which have increased by a factor of about 10 since 1900 to an enormous 5.3×10^9 tonnes in 1980. Deforestation adds about another 1×10^9 tonnes per annum. This must be considered in relation to the total atmospheric content of CO_2 which is about 720×10^9 tonnes. The various components of the overall global balance of carbon dioxide are not well quantified. CO_2 is removed from the atmosphere by photosynthesis in plants and is released in the natural processes of respiration and decay. These processes are roughly in balance. The oceans contain vast amount of CO_2 in inorganic form as well as in association with living organisms. Exchange of gas between the atmosphere and the upper layers of the ocean is rapid. In some areas there may be net release of CO_2 and in other areas net removal. Overall the oceans represent a net sink for CO_2. If all the anthropogenic emissions remained in the atmosphere we should expect to be seeing an increase in the CO_2 level by about 0.75% of its value each year. The measured CO_2 levels are increasing at about half this rate although the upward trend is accelerating with increasing emissions. Before the start of the industrial revolution it is estimated that the mean atmospheric concentration was below 300 p.p.m. From 1958 to 1988 the records at Mauna Loa Observatory in Hawaii show an increase from 315 to 350 p.p.m in the annual average concentration.

Global Warming

The rates of growth in the atmospheric levels of the other greenhouse gases and their contribution to warming relative to CO_2 are shown in Table 2 and Fig. 2.3 (Ramanathan *et al*). Molecule for molecule, changes in CH_4, N_2O, and the CFCs have more effect than changes in CO_2. Fig.2.3. indicates the theoretical global warming at

Table 2.2 Atmospheric levels and rates of change of the greenhouse gases used to calculate global warming

Gas	Global average mixing ratio (p.p.b) 1980	2030	Rate of rise
Carbon dioxide	339000	450000	2.4% yr^{-1} in emissions
Methane	1650	2340	0.7% yr^{-1}
Nitrous oxide	300	375	0.45% yr^{-1}
CFC 11 ($CFCl_3$)	0.18	1.1	3% yr^{-1}
CFC 12 (CF_2Cl_2)	0.28	1.8	3% yr^{-1}
Ozone (tropospheric)			0.23% yr^{-1}
Ozone (stratospheric) assumed to be +3.8% at 10 km and – 38% at 40 km			

Source: 'Stratospheric Ozone', UK Stratospheric Ozone Review Group, HMSO, London, 1987

equilibrium for given rates of growth of the individual gases. The calculation takes no account of the years that would be necessary for the earth to make the transition to the new equilibrium state. Warming of the oceans, for example, occurs relatively slowly. Although CO_2 is the most important contributor the other gases taken together contribute about half the overall temperature rise. The situation with ozone is complex. There appears to be a gradual increase in the level of tropospheric ozone due to emissions of NO_x, hydrocarbons, *etc.*, at the same time as a reduction in stratospheric ozone. The surface warming effect depends on the vertical distribution of changes in ozone concentration (Table 2.2).

Although there is general agreement that we are committed to an increase in mean global temperature the climatological consequences of this are less well understood. Modelling the effect of increased greenhouse gas levels on the global climate is an enormously complex problem requiring the most advanced computers available. Three-dimensional global circulation models describe the verticle, latitudinal, and longitudinal variations in conditions and attempt to compare the present situation with that when for example the CO_2 level has doubled. Changes in surface and

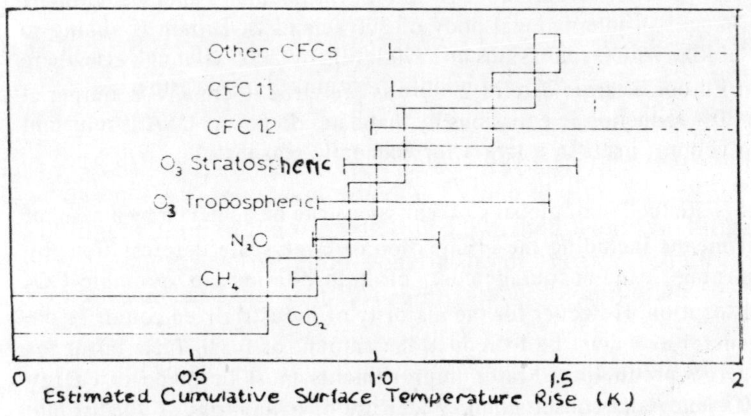

Fig.2 3. *Equilibrium surface warming by 2030 AD calculated by Ramanathan* et al. *on the basis of the assumed concentrations and rates of increase of trace gases listed in Table 2.2. The bars indicate uncertainties associated with the projections of release rates.*

atmospheric temperatures, cloud cover, evaporation, and precipitation, *etc.* all follow the changed radiation balance but the effects are not equally distributed over the globe. For CO_2 doubling most models now suggest an increase of 2—3 °C in mean global temperatures with the largest increases of 8—10 °C over N. Europe and N. Asia (> 50°N) in winter and increases of up to 4 °C in Antarctica. Although the global precipitation might increase by 5—10%, the studies suggest that the tropics and areas bordering the eastern coasts of continents would become generally wetter and the sub-tropical regions become drier. This might be critical for some central African regions which already suffer severe drought conditions. Major changes in sea level arising from the melting of the Antarctic ice-cap seem possible only on a time scale of several centuries. A significant thinning of the Arctic ice seems to have occurred already and is currently being studied. The last decade has been the hottest since accurate records began to be taken in the mid-nineteenth century. As yet there is no proof that this is due to the greenhouse effect and it may be many years before this can be established unequivocally.

International discussions are already taking place with a view to limiting the emissions of greenhouse gases. The second World Climate Conference met in Geneva in 1990. It had as its technical basis a report from the UN Intergovernmental Panel on Climate Change, an international body of 300 scientists. Britain is aiming to stabilize its CO_2 emissions at 1990 levels by 2005 although elsewhere in Europe a target date of 2000 was preferred. Germany is aiming at a 30% reduction in emissions by that date. So far the USA is reluctant to commit itself to a target for economic reasons.

Reduction of global CO_2 emissions can be achieved by a number of means including the elimination of large scale deforestation by burning and encouragement of reforestation to promote CO_2 absorption. However for the majority of industrialized countries the general aim must be to reduce the amount of fossil fuels burnt for energy production. Major improvements must be made in energy efficiency and conservation. Wider use of natural gas as a substitute for coal will result in some benefit since the mass of CO_2 emitted per unit of heat released is less: gas 0.43, oil 0.62, coal 0.75 ktonne MW_{yr}^{-1}. The other alternatives are the use of renewable energy sources such as wind, solar, wave, and tidal power or the further development of nuclear energy. The last of these options seems unlikely in the short term for both economic and environmental reasons.

DEPLETION OF STRATOSPHERIC OZONE

The Ozone Layer

Although ozone occurs in the troposphere and plays an important role in air pollution chemistry, about 90% of the total ozone content of the atmosphere occurs in the stratosphere at altitudes between 15 and 50 km. The ozone layer acts as a filter for ultra-violet radiation from the sun, removing most of the radiation below 300 nm. This serves to protect humans from the adverse effects of UV which become significant below 320 nm — decreasing wavelength corresponds to higher energy photons which can cause sunburn and types of skin cancer. Any depletion of stratospheric ozone would therefore lead to a larger amount of UV radiation incident on the earth's surface and an increased risk of the induction of cancers.

Concern was first expressed about this risk in the early 1970s in connection with emissions of nitrogen oxides from supersonic aircraft such as Concorde, which fly in the lower stratosphere. The nitrogen oxides are potential catalysts for the destruction of ozone. This particular effect is now thought to be relatively minor and attention has been switched to halogen compounds, especially the CFCs or freons. The freons are a group of chlorofluorocarbons used as aerosol propellants, refrigerants, and as gases for the production of foamed plastics. Their attraction lies in the fact that they are non-toxic, non-flammable, and chemically inert. Global production of the two commonest gases, CFC 11 ($CFCl_3$) and CFC 12 (CF_2Cl_2), rose rapidly from below 50 000 tonnes per annum in 1950 to 725 000 tonnes per annum by 1976 decreasing slightly to 650 000 tonnes in 1985. About 90% of this is released directly to the atmosphere while the remainder, representing the refrigerant use, will be released when the equipment containing it is eventually discarded.

The levels of CFCs in the atmosphere are extremely small, (less than 1 p.p.b., *see* Table 2.2), but are rising at a rate which correlates well with the known emissions. They are resistant to attack by molecules, radicals, or the UV radiation present in the troposphere and not subject to significant dry deposition or rain-out. The higher energy UV radiation in the stratosphere can however lead to photodissociation forming chlorine atoms which can in turn lead to the destruction of ozone. Despite the slow exchange of air between the troposphere and the stratosphere this effect is now known to be significant.

Ozone Depletion

The chemistry of ozone depletion is complex but a basic outline of the important processes is as follow. The ozone is formed from the dissociation of molecular oxygen by short wavelengtn UV radiation in the upper stratosphere:

$$O_2 \xrightarrow{UV} O^\bullet + O^\bullet \quad (1)$$

$$O^\bullet + O_2 + M \rightarrow O_3 + M \quad (M = \text{inert third body}) \quad (2)$$

However, ozone itself is rapidly photodissociated:

$$O_3 \xrightarrow{UV} O_2 + O^\bullet \quad (3)$$

so the so-called 'odd oxygen' species, O_3 and O, may interconvert many times before they destroy one another by:

$$O^\bullet + O_3 \rightarrow O_2 + O_2 \quad (4)$$

In fact, measurements of the ozone profile in the atmosphere suggest that ozone destruction must be considerably faster than could be achieved by reaction (4) alone. These other mechanisms can be represented by

$$X + O_3 \rightarrow XO + O_2 \quad (5)$$

$$XO + O^\bullet \rightarrow X + O_2 \quad (6)$$

Net effect $O^\bullet + O_3 \rightarrow O_2 + O_2$

If X = NO, the reactions form and destroy NO_2; if $X = Cl^\bullet$, the reactions form and destroy ClO. $X = OH^\bullet$ gives a third ozone destruction cycle. Other sets of reactions involving NO and $Cl^\bullet$ simply achieve the interconversion of O_3 and $O^\bullet$ and therefore have no effect on the net ozone levels. The reactive NO_x and $Cl^\bullet$ species can be removed by the formation of the relatively stable 'reservoir' molecules HNO_3 and HCl o. the somewhat shorter lived chlorine nitrate $ClONO_2$. About half the stratospheric content of NO_x is stored as HNO_3 and about 70% of the chlorine as HCl. Although these may be reactivated by conversion back to NO_x and $Cl^\bullet$, they may eventually be transferred back to the troposphere and removed to the ground by rain-out. N_2O is significant as it photolyses in the stratosphere and forms the major precursor of NO_x. Methane reactions produce hydrogen containing species including water vapour and HCl.

The Antarctic Ozone 'Hole'

This was the general picture (although a highly simplified account) of the homogeneous stratospheric chemistry as understood before 1985. Even greater depletions were subsequently observed in 1987 and 1989. In fact in the lower stratosphere 97% of the ozone in the range 14—18 km was lost. Subsequent aerial surveys and analysis of satellite data have confirmed this phenomenon and led to a complete reappraisal of the chemistry and meteorology involved.

During the dark, cold Antarctic winter upper stratospheric air moving from low to high latitudes subsides and as it does so develops a strong westerly circulation pattern. This produces a vortex which effectively isolates the air in the lower stratosphere over the Antarctic continent from the air at lower latitudes. Within the vortex the temperature falls progressively until below about —80 °C polar stratospheric clouds (PSCs) may form. These are composed of very small particles (1 μm) of nitric acid trihydrate ($HNO_3.3H_2O$). A further drop in temperature of about 5 °C may result in water ice crystals being formed. These are rather larger (10 μm). It is the heterogeneous reactions involving these cloud crystals which dramatically alter the chemistry.

Basically these reactions convert chlorine from its inactive, reservoir forms ($HCl, ClONO_2$) into forms which are active ozone depletors ($Cl^{\bullet}$, ClO). HCl is readily incorporated into ice crystals and can undergo reaction with $ClONO_2$.

$$\underset{\text{ice}}{HCl} + \underset{\text{gas}}{ClONO_2} \rightarrow \underset{\text{gas}}{Cl_2} + \underset{\text{ice}}{HNO_3} \qquad (7)$$

and

$$H_2O + ClONO_2 \rightarrow HOCl + HNO_3 \qquad (8)$$

The nitric acid is left in the ice phase. The chlorine remains in the gas phase until the polar spring when the sun reappears photodissociates it to chlorine atoms:

$$Cl_2 + h\nu \rightarrow Cl^{\bullet} + Cl^{\bullet} \qquad (9)$$

The $Cl^{\bullet}$ atoms rapidly react with ozone generating ClO:

$$Cl^{\bullet} + O_3 \rightarrow ClO + O_2 \qquad (10)$$

In the winter the stratosphere is thus chemically 'preconditioned' by heterogeneous reactions so that in the spring very rapid ozone depletion occurs. In addition to the ozone destruction cycle represented by equations (5) and (6) with $X = Cl^{\bullet}$ it is now recognized that chlorine monoxide dimers are also important. This was realized because the oxygen atom concentrations in the lower stratosphere are too low to account for the observed ozone destruction rates:

$$ClO + ClO + M \rightarrow Cl_2O_2 + M \qquad (11)$$

$$Cl_2O_2 \xrightarrow{UV} Cl^{\bullet} + ClOO^{\bullet} \qquad (12)$$

$$ClOO^{\bullet} + M \rightarrow Cl^{\bullet} + O_2 + M \qquad (13)$$

These reactions by-pass the $ClO + O^{\circ}$ reaction as a route for reconversion of ClO back to Cl. Reactions of bromine atoms in addition to those of chlorine atoms are now thought to account for about 20% of the ozone depletion. Bromine emissions occur in the form of methyl bromide which has natural and man-made sources plus another family of halocarbons, the Halons—Halon 1301 is CF_3Br, Halon 1211 is CF_2BrCl (Table 2.3).

But is the 'hole' in the Antarctic ozone layer significant for the rest of the globe? Several facts suggest that it is. As the Antarctic spring progresses the vortex breaks up and the ozone depleted air can then be transported to lower latitudes. Such an event has been observed over Australia. In the Northern Hemisphere airborne studies of the Arctic winter were carried out it 1988/9. Although the temperatures are not as low as in the Antarctic and the occurrence of stratospheric clouds not as common, nevertheless they are formed and the existence of a similar chemistry with high ClO concentrations has been demonstrated. The extent of ozone depletion is less marked—perhaps 10—25% in the range 20—25 km altitude representing a reduction of some 3% in total column ozone. Data from the network of ground level monitors for column ozone show a clearly decreasing trend in the Northern Hemisphere since 1970 although the considerable variability in the data makes the precise percentage decrease sensitive to the start date assumed. For Europe and North America the decrease is 2.5 to 3.5% per decade with an indication that the trend has accelerated in the last decade, in parallel with the worsening conditions in the Antarctic.

International Control Measures

The UN Convention on the Protection of the Ozone Layer (the 'Vienna Convention') was agreed in 1985 and subsequently measurers

Table 2.3 Atmospheric lifetimes, emissions, and ozone depletion potentials for halogenated compounds

Compound *name*	Chemical formula	*Atmospheric lifetime Year*	1985 emissions kt yr^{-1}	*Ozone depletion potential (CFC 11 = 1)*	Ozone removal % *(1985 emissions)*
CFC 11	$CFCl_3$	77	281	1.0	30.4
CFC 12	CF_2Cl_2	139	370	1.0	40.0
CFC 13	$C_2F_3Cl_3$	92	138	0.8	11.7
Carbon tetrachloride	CCl_4	76	66	1.06	7.6
Methyl chloroform	CH_3CCl_3	8.3	474	0.10	5.1
HCFC 22	CHF_2Cl	22	72	0.05	0.4
HFC 134a	$C_2H_2F_4$	10	0	0	0
Halon 1301	CF_3Br	101	3	11.4	3.7
Halon 1211	CF_2BrCl	12.5	3	2.7	0.9

Source: 'Stratospheric Ozone', UK Stratospheric Ozone Review Group, HMSO, London, 1988

to reduce the emissions of various halocarbons were incorporated into the 'Montreal Protocol' in September 1987. A meeting in London in June 1990 further tightened the restrictions under the protocol. More than 70 countries are now signatories. The use of both CFCs and Halons will be phased out by the year 2000 as will the use of carbon tetrachloride,CCl_4. Methyl chloroform will be phased out by 2005. In the immediate future replacement chemicals such as the HCFCs will be used. Because these contain hydrogen atoms as well as halogens they are more reactive in the troposphere and have shorter lifetimes. Their potential impact on the stratosphere is therefore much reduced. Even so the expectation is that the total stratospheric chlorine loading will peak after 2000 at about 4 p.p.b. and only decline slowly through the remainder of the twenty first century. How long it takes to fall to the pre-war level of below 1 p.p.b. depends on the extent of global participation in implementing restrictions and the extent of future use of alternative chlorine containing compounds such as the HCFCs (Table 2.3).

REFERENCES

The Greenhouse Effect, Climate Change, and Ecosystems', (SCOPE 29), ed. F Bolin, B.R. Doos, J. Jager, and R.A. Warwick, John Wiley and Sons, Chichester, 1986.

The Greenhouse Effect and Terrestrial Ecosystems of the UK', ITE Research Publication No. 4, ed. M.G.R. Cannell and M.D. Hooper, HMSO, London, 1990.

Climate Change, The IPCC Scientific Assessment', ed. J.T. Houghton, G.J. Jenkins, and J.J. Ephraums, Cambridge University Press, Cambridge, 1990.

V. Ramanathan, R.J. Cicerone, H.B. Singh, and J.T. Kiehl, *J. Geophys. Res.* 1985, **90,D3**, 5547.

UK Review Group on Stratospheric Ozone,'Stratospheric Ozone', HMSO, London, 1987.

UK Review Group on Stratospheric Ozone,'Stratospheric Ozone', HMSO, London, 1988.

UK Review Group on Stratospheric Ozone,'Stratospheric Ozone', HMSO, London, 1990.

J.C. Farman, B.J. Gardiner, and J.D. Shanklin, *Nature (London)*, 1985, 315, 207.

Department of the Environment,'Digest of Environmental Protection and Water Statistics', No. 13, HMSO, London, 1990.

R.C. Flagan and J.H. Seinfeld, ' Fundamentals of Air Pollution Engineering', Prentice Hall, Englewood Cliffs, 1988.

CHAPTER 3

Emissions to Atmosphere

The primary components of pure dry air are nitrogen, N_2 (78.1%), oxygen, O_2 (20.9%), argon, Ar (0.9%), and carbon dioxide, CO_2 (0.035% or 350 p.p.m.). Water vapour is present in amounts which typically range from 0.5-3% at ground level, depending on temperature and relative humidity. Analysis of air samples reveals the presence of hundreds of other substances in trace amounts. Some of these can be explained directly in terms of their emissions either from natural sources or human activity. Others must have arisen indirectly by chemical processes in the atmosphere. We distinguish these as *primary* and *secondary* pollutants. The secondary pollutants, including gases like ozone and particulate compounds such as sulphates, are dealt with in chapter 5. Here we concentrate on the primary pollutants which arisen from sources we can usually identify.

NATURAL EMISSIONS

Sulphur Species

Within a local area the levels of most pollutants are usually dominated by the contributions for which man himself is responsible. However nature is also a generator of what we call 'pollutants' and on a global scale the natural emissions may far outweigh human emissions. This is illustrated in Table 3.1. Sulphur in the form of SO_2 and some H_2S is emitted most dramatically from volcanoes. However much large amounts are emitted within sea spray (as sulphate) and from biological processes. In the absence of air, biological decay results in emissions of hydrogen sulphide, H_2S, and organic compounds such as dimethyl sulphide. Both terrestrial and oceanic emissions are significant. The grouping together of all sulphur compounds is appropriate since H_2S and organic sulphur compounds are converted to SO_2 in the atmosphere. Various estimates place the total emissions at 150-250 Tg sulphur per year (1 Tg = 1 million tonne). Since the global SO_2 from combustion emissions is 70–100 Tg S yr^{-1}, the natural sulphur emissions are greater than the anthropogenic emissions.

Nitrogen Species

Biological processes in soil lead to the release of all of the common nitrogen oxides, NO, NO_2, and N_2O. The amounts involved are very uncertain but for NO and NO_2 are of the order of 10 Tg N yr^{-1}. Lightning and biomass burning are other major sources. Oxidation of NH_3 to NO occurs in the troposphere and some nitrogen in the form of HNO_3 is transferred to the troposphere from the stratosphere. These sources total 20–30 Tg N yr^{-1}. In comparison the anthropogenic emissions of NO + NO_2 from combustion are about 20 Tg N yr^{-1}. The sources of nitrous oxide, N_2O are not as well understood as the main loss mechanism which is decomposition in the stratosphere (6–10 Tg N_2O yr^{-1}) The major source is the release from soil especially in situations where

TABLE 3.1 Natural emissions of S and N Compounds

Source	*Tg S yr*$^{-1}$	Source	*Tg N yr*$^{-1}$
Volcanoes	< 2	Lightning	8
Biogenic gases from land	35	NH_3 oxidation	1–10
Biogenic gases from water	35	From stratosphere	0.5
Sea spray	171	Biogenic production	8
		Biomass burning	12
Natural total	243	Natural total	33
Anthropogenic	75	Anthropogenic	21

Source: *D. Möller Atoms. Environ. 1984, 18, 19 and 29.*
J.R. Logan, J. Geophys. Res. 1983, 88, 10785.

fertilizer has been added. smaller releases occur from the oceans and from combustion processes.

Anthropogenic emissions of ammonia, NH_3, the other nitrogen-containing gas, are only a few Tg yr^{-1}, mainly from waste treatment. Natural emissions from biological decay and animal excrement exceed 100 Tg yr^{-1} although the exact amount is highly uncertain.

Hydrocarbons

The largest natural sources of methane are anaerobic fermentation of organic material in rice paddies and in northern wetlands and tundra, plus enteric fermentation in the digestive systems of ruminants (*e.g.* cows). Methane is also released from insects, from coal mining, gas extraction, and biomass burning. Total emissions are 300–550 Tg yr^{-1} and appear to the increasing at a rate of 50 Tg yr^{-1}.

Heavier hydrocarbons, such as terpenes, are released directly to the atmosphere from trees. Global emissions considerably exceed anthropogenic emissions.

STATIONARY COMBUSTION SOURCES

Carbon Dioxide

The main elemental components of fossil fuels are carbon and hydrogen. When burnt with the oxygen present in air, carbon is converted to carbon dioxide, CO_2, and hydrogen to water vapour, H_2O. It is these overall reactions which convert the chemical energy of the fuel into useful heat. CO_2 is harmless to man at concentrations below about 1% being a natural product of our metabolic processes which is exhaled in our breath. The concern over its wider role in the atmosphere and its possible effect on climate has already been discussed.

Carbon Monoxide

In a flame the hydrocarbon fuel molecules are progressively broken down to smaller fragments and combined with oxygen. The end product of this stage of the reaction is carbon monoxide, CO. subsequent further oxidation converts the carbon monoxide to CO_2. Providing there is sufficient oxygen present, all the CO should be converted to CO_2. However, the formation of carbon monoxide is relatively fast whereas its removal by the reaction:

$$CO + OH^\bullet \rightarrow CO_2 + H^\bullet$$

is relatively slow. For its removal to be complete requires high temperatures to maintain rapid reaction and sufficient time before the gases are cooled. These conditions are met fairly readily in most combustion appliances providing they are properly adjusted. Faulty or improperly adjusted appliances can produce dangerous amounts of CO (several percent in the flue gas) usually due to some abnormal limitations of the air supply. CO emissions from internal combustion engines are more of a problem and are discussed below.

Soot

Formation of soot commonly accompanies carbon monoxide for–mation and is generally due to inadequate air supply. In these circumstances partially degraded fuel molecules can polymerize to produce carbon nuclei which can accumulate further material by surface reactions and coagulation. The resulting soot particles are commonly sub-micron (<1 μ m) and because they are of comparable size to the wavelength of light they are effective both at light absorption and light scattering. A relatively small mass concentration wil! therefore render the exhaust or flue gases opaque.

Combustion of fuel oils achieved after spray atomization. Droplets of lighter oils (e.g. diesel, kerosine, gas oil) will normally evaporate completely and any soot that is formed will arise from the gas-to-particle route just described. With heavy fuel oils only the lighter components are rapidly vaporized, leaving a tarry residue which is thermally cracked by the heat of the flame. The end product is a hollow, spherical, carbonaceous particle not a lot smaller than the original oil droplet. Given sufficient time and high enough temperatures, these will burn away but some may survive and be deposited within the combustion system or emitted from the chimney. This type of soot has a particle size up to 50–100 μm. Since heavy fuel oils contain sulfur, some of which may be converted to sulfuric acid, an additional problem can be the formation of acid smuts. Soot deposited on the inside of the chimney wall can become saturated with condensing acid. Flakes of this material later blown out of the chimney will eat holes in clothing and corrode the paintwork of cars. This is a very local problem restricted to a few hundred metres distance from the chimney.

Combustion of coal proceeds in a rather similar way to oil droplets. Volatile material is released and burns as a gas, the remaining char burns more slowly and a final residue of ash is left. Combustion of pulverized coal suspended in air can normally be controlled to avoid smoke formation. However burning lumps of coal on a great is more of a problem because the mixing of air with the volatile matter released by the heat is rather inefficient. Anthracite, coke and the other manufactured smokeless fuels contain low levels of volatile material and so avoid the problem. Incomplete combustion of volatile matter is also the major problem with wood smoke and smoke from bonfires or stubble burning.

Hydrocarbons

Most boilers and central hearing units burning fossil fuels result in very low emissions of gaseous hydrocarbons or oxygenated hydrocarbons such as aldehydes. The term Volatile Organic Compounds (VOCs) is used to describe organic material in the vapour phase excluding methane. These compounds are of importance in internal combustion engines.

High molecular weight hydrocarbons are often found in association with soot. Because of the relatively low temperatures on a domestic grate, the smoke contains not only soot but also tarry material that has essentially been distilled from the coal and recondensed in the flue. Such material contains potentially toxic compounds. Workers in the old types of coal gasification plant before the advent of modern standards of environmental control were susceptible to a type of cancer that was attributed to polynuclear aromatic hydrocarbons, referred to as PNAs or PAHs of which benzo (*a*) pyrene is one example:

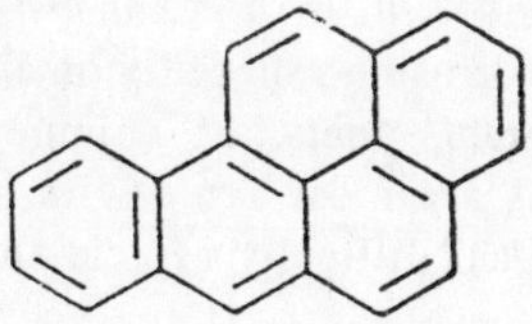

Consequently there has been a long-standing interest in the presence of such carcinogens in the general environment and in the levels emitted from combustion processes. Attention has been switched from coal smoke to diesel engine exhaust smoke as a source of these compounds but it remains to be established whether there is a significant health hazard at the very low concentrations to which the general public are exposed.

Ash

Bituminous coals burnt in power stations tend to be of lower quality; a typical ash content might be 16%. Coke and anthracite are rather cleaner fuels, with ash contents below 10%. In the U K emission of pulverized fuel ash from current power stations is limited to 115 mg m^{-3} and this requires the removal of over 99% of the ash originally present in the flue gases. Electrostatic precipitators are the usualmeans of achieving the high particle removal efficiency.

Sulphur Oxides

Sulphur dioxide, SO_2 arises from the sulphur present in most fuels. In coal the sulphur takes the form partly of mineral matter (pyrites, FeS_2) and partly of organic sulphur compounds. Much of the sulphur in the lighter petroleum fractions is removed during refining but that concentrated in the heavier fuel oils is difficult to remove being strongly bound in aromatic ring compounds. During combustion conversion of sulphur dioxide is virtually complete although about 10% may be retained by coal ash. For the highest sulphur fuel oils (3%S) the flue gas concentrations can reach 2000 p.p.m., while for a typical power station coal of 1.6% S the concentration is about 1200 p.p.m. No control over SO_2 emissions can be achieved by modification of combustion conditions and reductions must be sought by pre-treatment of the fuel or desulphurization of the flue gases after combustion.

A small fraction of the SO_2, usually less than 1%, may be further oxidized to sulphur trioxide, SO_3 which combines with water vapour to form sulphuric acid H_2SO_4 in the flue gas. This can have serious implications for in-plant corrosion especially on any low-temperature surfaces in the air heaters, ducts, or chimney where the acid condenses. The problems are most acute with heavy fuel oils for which there is the additional difficulty of acid smuts.

Nitrogen Oxides

There are two main routes to the formation of the nitrogen oxides NO and NO_2 together described as NO_x. One involves the combination of nitrogen and oxygen in the air at the peak flame temperatures to form nitric oxide, NO. This is termed 'thermal NO_x'. The other starts with the nitrogen originally present in the fuel. Depending on the conditions, some of the fuel-nitrogen will be converted to NO and some to nitrogen gas, N_2. Factors such as the burner design, the intensity of combustion, the overall shape and size of the furnace, and the amount of excess air all influence NO formation and can be modified to achieve a certain measure of control. However this falls considerably short of eliminating the emissions. Typical flue gas concentrations for coal-fired power stations are about 550 p.p.m., but with new designs of low-NO_x burners currently being installed this will be reduced by about 40%. Nitrogen dioxide, NO_2 forms only a small fraction of the waste gases from combustion (usually less than 10%) so the description 'NO_x emissions' actually refers mainly to NO emissions. Once in the atmosphere, oxidation of NO to NO_2 occurs and the relative proportions of the two oxides may then be comparable.

Hydrogen Chloride

In addition to SO_2 and NO, the combustion of coal produces hydrogen chloride, HCl originating from chlorides in the fuel at a level of about 0.2% by weight. Although not generally considered to be as important a contributor to pollution problems, it adds to the acidity of precipitation. Incineration of waste containing chlorinated plastics such as PVC also produces HCl. Petroleum products have no naturally occurring chlorine but leaded petrol contains both chlorine and bromine added in the form of ethylene dichloride and ethylene dibromide in conjunction with the lead anti-knock additives. The chlorine and bromine scavenges the lead oxide formed during combustion and converts it to more volatile lead halides which are emitted from the exhaust. The remainder will be emitted as HCl and HBr but the total amounts involved are small compared to the HCl from coal combustion.

EMISSIONS FROM ENGINES

Gases

Internal combustion engines, whether of the spark ignition (petrol) or compression ignition (diesel) variety, have particular features which warrant discussion beyond that already given. The combustion takes place under high pressure instead of atmospheric pressure and the peak temperatures are higher than in a normal boiler. The time available for combustion is limited by the engine's cycle to a few milliseconds instead of a second or more.

In petrol engines, lacking any control devices, incomplete burnout of the fuel leads to high carbon monoxide and significant hydrocarbon emissions, especially during idling and deceleration. Nitrogen oxide emissions are also high due to the high temperatures but are at a maximum during acceleration. Disel engines have much lower carbon monoxide and hydrocarbon emissions but comparable NO_x emissions to a petrol engine.

Particulates

Smoke formation is not a significant problem with spark ignition engines because the fuel and air are well mixed before entering the cylinders. Diesel engines suffer from the disadvantage of producing more smoke especially under heavy load or acceleration conditions. The reason is the relatively poor mixing of air with the fuel spray injected into the cylinder. This produces regions which are too rich in fuel for complete combustion, leading to soot formation. Cities with large numbers of diesel vehicles such as buses suffer from elevated roadside levels of smoke compared to the general urban environment. The odours associated with diesel emissions arise from partially burnt, oxygenated hydrocarbons such as aldehydes.

Lead

Emissions of lead from automobile exhausts arise from the use of lead tetra-alkyl anti-knock additives to improve the octane rating of the petrol. About 70% of the lead is emitted from the tail-pipe, mainly as mixed halides. Increasing concern over the potentially harmful effects of lead on health, particularly the health of children, resulted in a gradual decrease in the maximum permitted amount which can be added to petrol.

NON-COMBUSTION SOURCES

Gases

For the primary pollutants discussed so far, the emissions from combustion processes generally exceed those from other industrial sources although in local areas emissions from specific industries may be significant. Sulphur dioxide is emitted from sulphuric acid plants and from the roasting of sulphide ores during non-ferrous smelting. NO_2 is emitted from nitric acid plants and hydrocarbons or hydrogen sulphide from refineries. However the largest source of VOCs is the use of solvents, including those released from paints, together with the evaporative losses of gasoline during storage and distribution.

Other pollutants not encountered in combustion can be significant in the process industries. For example, hydrogen fluoride is emitted from brick kilns where fluoride-containing clays are used.

It is also emitted in primary aluminium smelting due to the fluoride in the molten electrolyte, from fertilizer works employing fluorapatite–a mixed calcium fluoride/phosphate ore, and from some glass-making processes. Fluoride pollution can cause fluorosis in cattle grazing in the vicinity of the source leading to bone damage and ultimately to death.

Dusts

Probably the most prevalent form of air pollution not necessarily connected with combustion is dust. Particulate matter other than soot can be emitted in many forms–pulverized fuel ash, metal oxide fumes, silica, *etc.* It is convenient to distinguish between different sizes of particles, the following terminology is commonly used :

Grit – large particles (>76μm diameter) which will rapidly settle out of the air due to gravity and are just visible to the naked eye.

Fume– very small particles (<1 μm diameter) which can remain suspended in air for long periods and which are visible only by electron microscopy.

Dust– the intermediate size range (1 μm < diameter < 76 μm) which can be seen under an ordinary optical microscope.

In addition to process emissions, dust can obviously arise by the disturbing action of outdoor industrial activity on the ground or on raw materials. Quarrying, open cast mining, tipping, digging, the action of heavy lorries, or simply a strong wind acting on stock piles can lead to grit and dust blowing beyond the site and causing a nuisance to others in the area. Such 'fugitive' emissions are difficult to quantify, and control depends on careful industrial practice with an awareness of the possible problems. Water can often be used to good effect to minimize the amounts of dust raised.

Odours

The last air pollutant is the most difficult to deal with scientifically because, to some extent, it is a subjective problem. The unpleasantness of an odour will be judged differently by different people and the sensitivity of individuals to the same level of odour is quite variable. Odours are frequently complex mixtures of airborne substances and may be near the limits of detection of current measurement techniques. One method of quantification is to determine the number of times a sample of odorous air has to be diluted with clean odour-free air so that 50 percent of a group of panellists can no longer detect it.

The most commonly encountered sources of odour arise from agricultural practices–intensive livestock production, disposal of slurry to land, *etc*. Of more immediate concern to town-dwellers are sewage works and landfill sites. Better control of tipping and coverage of the waste with inert material has reduced the problems at such sites. Perhaps the most serious smells arise with the rendering of animal wastes.

REFERENCES

J. B. Heywood, 'Internal Combustion Engine Fundamentals', McGraw Hill, London, 1988.

Directive 88/76/EEC, 'Emissions standards for vehicles up to 3.5 tonnes'. *Off. J. Eur. Comm.*, 1988 L36, 1.

Directive 88/436/EEC.'Limiting gaseous pollutants from diesel vehicles up to 3.5 tonnes', *Off. J Eur. Comm.*, 1988, L214,1.

Technical Data on Fuel', 7th Edn., ed. J.W. Rose and J.R. Cooper, Scottish Academic Press, Edinburgh, 1977.

CHAPTER 4

Air Quality And Emissions

Smoke and Sulphur Dioxide

Following the outcry over the London smog of 1952 and the Clean Air Act of 1956, a concerted effort was made to reduce levels of smoke and sulfur dioxide in the air. The main components of this effort were control of visible smoke emissions from industrial chimneys, control of chimney heights to ensure adequate dispersion of SO_2, and the introduction of 'smoke control zones' by local authorities. Within such zones the burning of bituminous coal in domestic grates is forbidden and smokeless fuels', *viz.* anthracite, coke, or other manufactured smokeless fuels, must be used in approved types of grates or stoves. The alternative to solid fuels is to switch to electricity, oil, or gas, all of which have negligible smoke emissions compared to coal. Domestic coal is limited and the major users are now the electricity supply industry and steel making. Even this pattern is changing with the generating companies switching to gas for new power stations for economic and environmental reasons.

'Black Smoke' is related to the darkness of the stain on a filter paper through which air has been drawn. The result is reported as a mass concentration ($\mu g\ m^{-3}$) based on a calibration devised at a time when smoke was primarily coal smoke. Different smokes have different degrees of blackness per unit mass. For example diesel engine soot is 3 times balcker than coal smoke and this is incorporated into the estimation of the 'black smoke' emissions. It is estimated that over 50% of smoke readings in city centre areas may be contributed by diesel engine emissions although the overall national average is 36%.

Nitrogen Oxides

Emissions of nitrogen oxides have gradually increased due to increased traffic density. Since nitrogen dioxide is the main health hazard, most attention has been paid to monitoring urban levels of NO_2 as opposed to total NO_x.

Table 4.1 FEC air quality standards

Pollutant	*Measure*	*Limit values* ($\mu g\ m^{-3}$)			
		SO_2	...	*Smoke*	*Smoke*
SO_2/smoke	98th percentile	250	if	> 150	250
	of daily averages	350	if	< 150	
	Winter median	130	if	> 60	130
		180	if	< 60	
	Annual median	80	if	> 40	80
		120	if	< 40	
Lead	Annual average		2		
NO_2	98th percentile of hourly averages		200		

Carbon Monoxide

Emissions of carbon monoxide are predominantly from road vehicles and have increased as traffic density has increased. Carbon monoxide levels beside busy roads can under adverse conditions reach approximately half the occupational exposure level of 50 ppm but general urban levels are usually below 10 ppm and annual averages only 1—2 ppm 10 ppm is in fact the WHO guideline for 8 hour averages not one hour averages. Apart from roadside locations under low windspeed conditions, CO is most likely to be a problem in road tunnels and underground car parks where there may be limited ventilation.

Ozone

Ozone is a secondary pollutant formed in the atmosphere by photochemical reactions. However there is also a significant background level in the troposphere of about 30 ppb which arises partly from transfer from the stratosphere. The chemistry involved is discussed in chapter 5.

Metals

The EC air quality standard is $2\mu g\,m^{-3}$ annual average. Occupational exposure limits are valuable as a general indicator of relative toxicity when considering atmospheric levels of pollutants. However, tolerable levels of exposure appropriate for the general public are obviously much lower than those for healthy workers.

ATMOSPHERIC TRANSPORT AND DISPERSION OF POLLUTANTS

Wind Speed and Direction

In general low wind speeds result in high pollutant concentrations and vice versa. If we imagine the wind blowing across the top of a chimney emitting smoke at a constant rate, the *volume* of air into which the smoke is emitted is directly proportional to the wind speed. The *concentration* of smoke in the air is thus inversely proportional to the wind speed. A similar description can be applied to a source distributed uniformly over a wide area (*e.g.* domestic emissions from a city), Fig. 4.1. The concentration of pollutants in the hypothetical box of air into which the pollutants are mixed is proportional to the emissions rate and inversely proportional to the wind speed. In practice this picture is grossly oversimplified.

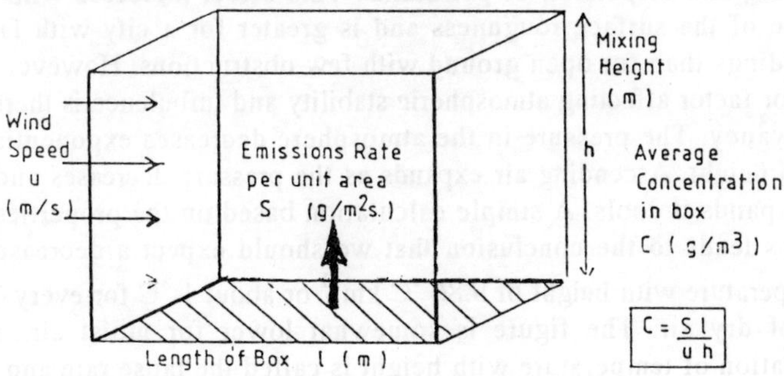

Fig. 4.1. *A simple box model applied to an area source such as a town*

The concentration of pollutants measured in urban areas rarely decreases with wind speed as rapidly as predicted.

For reasons which will be discussed later, the wind speed at ground level tends to drop overnight and rise again during the morning, especially during cloud-free conditions. Of course, emissions also tend to drop overnight—fewer fires, boilers, and furnaces are alight, fewer cars are on the roads. So some of the highest pollution levels occur in the morning when emissions increase rapidly before the wind speed picks up and dispersion conditions improve.

The people most affected by air pollution are those who are situated downwind of the major sources. A knowledge of the prevailing with direction is therefore important in predicting the likely impact of these sources. A more detailed presentation of the information can be made using a *wind rose* in which the length of the line is proportional to the frequency of occurrence of the wind in each sector of 45°or 30°. A similar approach can be used to show which wind directions are associated with the highest pollution levels.

The wind direction at a point is not sufficient to identify high pollution levels with the effect of distant sources. Over the scale of hundreds to thousands of kilometres the overall path of the particular mass of air must be computed for periods of perhaps 1 to 3 days.

ATMOSPHERIC STABILITY

The Lapse Rate

The roughness of the ground produces a certain amount of turbulence in the lowest layer of the atmosphere which promotes the mixing and dispersion of pollutants. This effect increases with the scale of the surface roughness and is greater for a city with large buildings than for open ground with few obstructions. However the major factor affecting atmospheric stability and turbulence is thermal buoyancy. The pressure in the atmosphere decreases exponentially with height. Ascending air expands as the pressure decreases and as it expands it cools. A simple calculation based on the properties of gases leads to the conclusion that we should expect a decrease in temperature with height of 9.86 °C km^{-1} or about 1 °C for every 100 m of dry air. The figure is somewhat lower for moist air. The variation of temperature with height is called the lapse rate and the calculation for the ideal case leads to what is known as the adiabatic lapse rate (a.l.r.).

In the real atmosphere the lapse rate can be greater than, smaller than, or close to the adiabatic lapse rate. The fundamentally affects the extent of vertical mixing of air.

Unstable situations occur with bright sunlight warming the ground to a temperature above that of the air. The air adjacent to the ground is subsequently warmed and rises due to its buoyancy. Such situations are common during daytime in summer, especially when the wind speed is low. High wind speeds tend to lead to *neutral* conditions with the lapse rate close to the adiabatic value.

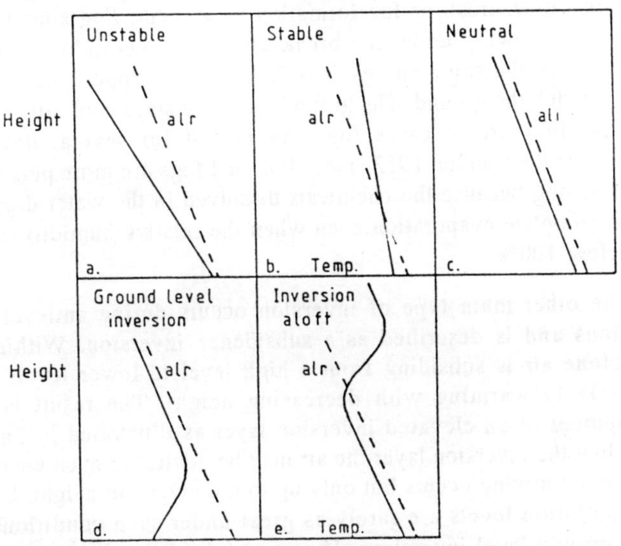

Fig. 4.2. *Schematic illustration of the atmospheric lapse rate for various stability categories. Full lines : actual temperature profile ; dashed lines : adiabatic lapse rate (a.l.r.)*

Temperature Inversions

Stable situations occur when the lowest layer of air is cooled by the ground beneath. The most common cause is overnight radiation cooling of the ground which often leads to ground level temperature inversions on clear nights. The temperature profile, which may have been shown in Figure 4.2.c in late afternoon, gradually changes into profile Figure 4.2.d overnight. The effect is that ground level emissions become trapped in the stable inversion layer which may not be more than 100—200 m deep. Emissions from tall industrial chimneys however may be above the inversion layer and be vertically dispersed by relatively good mixing conditions aloft. The following day the inversion layer is gradually eroded by the warming effect of the sun until, by mid-morning, the temperature profile has returned to that of neutral conditions (Figure 4.2.c) and the trapped pollutants are effectively released to be dispersed to higher levels.

Another factor contributing to high pollution levels during inversion conditions is the lowered wind speed. Since high level, faster moving air does not mix with low level air, there is no mechanism for the downward transport of momentum. The lowest level air therefore becomes stagnant. In the low temperature conditions, dew, frost, or fog formation may occur. Fog adds to the problem by slowing down the break-up of the overnight inversion layer because the sun's energy is reflected by its upper surface and does not reach the ground. The ground therefore stays cool rather than warming. In extreme cases fog may persist for several days as happened in the London 1952 smog. Polluted fogs are more persistent than clean fog because the chemicals dissolved in the water droplets prevent complete evaporation even when the relative humidity drops well below 100%.

The other main type of inversion occurs during anticyclonic conditions and is described as a subsidence inversion. Within an anticyclone air is subsiding from a high level to lower levels and progressively warming with decreasing height. The result is the development of an elevated inversion layer as illustrated in Figure 4.2. Below the inversion layer the air may be neutral or even unstable so that good mixing occurs but only up to the inversion height. Local urban pollution levels are rarely as great under such conditions as during ground level inversions. However subsidence inversions are often associated with warm, dry weather and they provide the ideal conditions for a long-range transport of pollution.

In the discussion so far, no reference has been made to the geographical situation in which air pollution levels are being considered. Towns situated in valleys are particularly susceptible to pollution problems. Cool air will tend to flow downhill into the valley so aggravating the problem of low level inversions. Mixing between the air in the valley and the air above is reduced. Fogs will persist longer. Often in winter a layer of polluted air over the two with cleaner air above can be clearly seen from the higher ground. Towns situated by the sea may be subject to sea breezes. The proximity of relatively warm ground to the cool water surface results in a circulation of air from sea to land. The sea breeze will be cooler than the overlying air, another example of inversion conditions, so in polluted areas conditions will be worse than for a corresponding inland site. Sea mists may blow inland aggravating the general discomfort.

DISPERSION FROM CHIMNEYS

Plume Rise

After the waste gases leave the top of a chimney the plume may rise a considerable height before achieving more or less horizontal travel. The major factor is the thermal buoyancy of the plume. The hotter the flue gases, the greater the plume rise. A lesser factor is the vertical momentum of the gases due to their efflux velocity out of the chimney top. Plume rise can mean that the effective chimney height is double the physical chimney height under low wind speed conditions and so is a considerable asset in achieving effective dispersion. Conversely any control technology for pollutant reduction which also reduces the flue gas temperature (such as a scrubber) results in poorer dispersion of the plume.

Ground Level Concentrations

The effluent gases leaving a chimney gradually entrain more and more air and the plume width both in the vertical and in the horizontal directions increases with downwind distance. In the horizontal direction there is, of course, a rapid fall-off in concentration as we move away from the plume centre line defined by the average and wind direction. The ground level concentration very close to the chimney will be zero because this point is well below the base of the plume. Some distance downwind the dispersing plume reaches ground level and the concentration rises rapidly to a maximum value. Thereafter the concentration falls off with increasing distance as the plume becomes more and more dilute Fig 4.3. Unstable conditions will bring the plume down to ground more rapidly, that is closer to the chimney, but the rate cf dilution is large so the concentration falls rapidly from its maximum value as we go to greater distances. Stable conditions result in the plume dispersing very slowly. It may remain visible for a considerable downwind distance. The point of maximum ground level concentration is also a long way from the chimney.

The basic models of plume dispersion suggest that the maximum ground level concentration will depend on the square of the chimney height. A 40% increase in chimney height should roughly halve the ground level impact. However this strictly refers to the effective chimney height including the plume rise so the actual degree of improvement may not be quite as great.

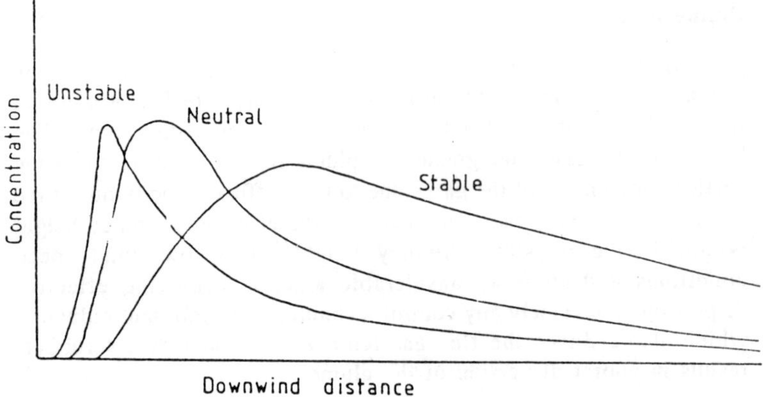

Fig. 4.3. *The variation in ground-level concentration along the plume centre line downwind of a chimney for various stability classes.*

Above the chimney top, the vertical dispersion of the plume may be hindered by an inversion layer. The pollutant are then trapped between the inversion height and the ground. At sufficient distance downwind, the concentration may be virtually constant at all heights within the mixing layer.

CHAPTER 5

Chemistry of Pollutants in The Lower Atmosphere

GASES

Atmospheric Photochemistry

Fig. 4.1 illustrates the processes which need to be considered between the emission of pollutants and their eventual destruction or removal to the ground. The chemical processes involved are complex and changes both in the gas phase and in the aqueous droplet phase are important. Not only do we have to consider the transformation of the primary pollutants but also the formation of secondary pollutants which can have adverse effects on the environment and on human health.

Of basic importance to an understanding of the gas phase chemistry is the effect of the sun. Photons of ultra-violet light provide a means of initiating chemical reactions which would otherwise not take place. In addition to stable molecules, photochemical reactions involve free radicals such as hydroxyl $OH^{\bullet}$ hydroperoxy $HO_2^{\bullet}$, and methyl $CH_3^{\bullet}$. Free radicals are extremely reactive and have very short lifetimes. Concentrations in the atmosphere are small but nonetheless significant. For example, $OH^{\bullet}$ concentrations in polluted atmospheres may be in the range $10^6 - 10^7$ radical cm^{-3}, *i.e.* one radical for every 10^{13} nitrogen molecules.

The overall process is one of oxidation, that is, the combination of atmospheric oxygen with the primary pollutants. For the three commonest inorganic pollutants the overall results are :

carbon monoxide	$CO \rightarrow CO_2$	carbon dioxide
nitrogen oxides	$NO, NO_2 \rightarrow HNO_3$	nitric acid
sulphur dioxide	$SO_2 \rightarrow H_2SO_4$	sulphuric acid.

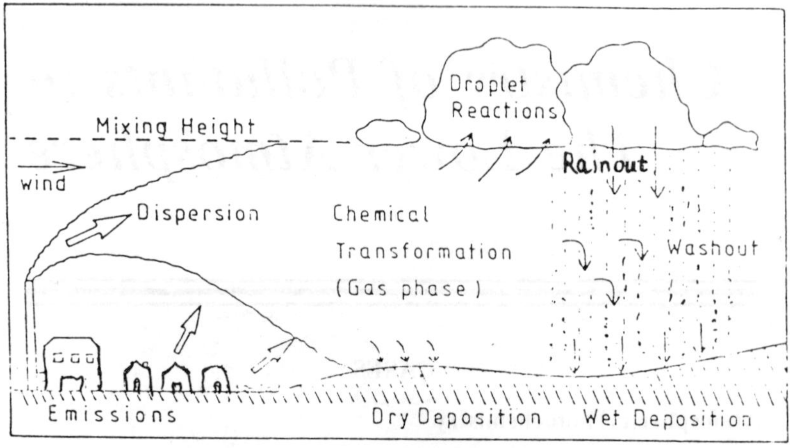

Fig. 5.1 Processes which may be involved between the emission of an air pollutant and its ultimate deposition to the ground.

Table 5.1. Chemical reactions for the atmospheric oxidation of CO, SO_2 and CH_4

Carbon monoxide

$$CO + OH^{\bullet} \rightarrow CO_2 + H^{\bullet}$$

$$H^{\bullet} + O_2 + M \rightarrow HO_2^{\bullet} + M \quad (M = \text{an inert third body})$$

Sulphur dioxide

$$SO_2 + OH^{\bullet} \rightarrow HSO_3^{\bullet}$$

$$HSO_3^{\bullet} + O_2 \rightarrow SO_3 + HO_2^{\bullet}$$

$$SO_3 + H_2O \rightarrow H_2SO_4$$

Methane

$$CH_4 + OH^{\bullet} \rightarrow CH_3^{\bullet} + H_2O$$

$$CH_3^{\bullet} + O_2 + M \rightarrow CH_3O_2^{\bullet} + M$$

$$CH_3O_2^{\bullet} + NO \rightarrow CH_3O + NO_2$$

$$CH_3O^{\bullet} + O_2 \rightarrow HCHO + HO_2$$

$HO_2^{\bullet}$ to $OH^{\bullet}$ conversion

$$HO_2^{\bullet} + NO \rightarrow OH^{\bullet} + NO_2$$

Later we shall consider the involvement of the two acids in particles and droplets but as far as the gas phase chemistry is concerned these species mark the end point of the process. For organic hydrocarbon species there may be a number of intermediate stable molecules formed but the overall process is rather like combustion with the end product being carbon monoxide ; for example :

methane CH_4 → formaldehyde HCHO → carbon monoxide

In all these cases the main species initiating the sequence of reactions is the hydroxyl radical :

$$CO + OH^{\bullet} \rightarrow CO_2 + H^{\bullet} \qquad (1)$$

$$NO_2 + OH^{\bullet} \rightarrow HNO_3 \qquad (2)$$

$$SO_2 + OH^{\bullet} \rightarrow HSO_3^{\bullet} \qquad (3)$$

$$CH_4 + OH^{\bullet} \rightarrow CH_3^{\bullet} + H_2O \qquad (4)$$

In the case of nitric acid formation, there are no remaining free radicals to continue the chain of reactions. In the other cases the hydroxyl radical is eventually regenerated but only after several further steps which interlink the chemistries of the various species (Table 5.1). The carbon monoxide oxidation is a slow process and the lifetime of CO in the atmosphere is several years. The oxidation rate of SO_2 can be around 2% h^{-1} in urban air or a factor of 10 lower in clean air resulting in an overall lifetime of a few days. Radicals other than $OH^{\bullet}$ such as $HO_2^{\bullet}$, $CH_3O_2^{\bullet}$, or other hydrocarbon peroxy radicals will also attack SO_2 but at a slower rate and their contribution to the overall oxidation is thought to be relatively small.

Where the hydroxyl radicals come from in the first place. One source present even in non-polluted atmospheres at a background level of 20-40 ppb. is ozone, O_3. Ultra-violet light of wavelength below 310 nm can dissociate ozone producing electronically excited oxygen atoms, O^* which rapidity split molecules of water vapour :

$$O_3 \xrightarrow{UV\ light} O_2 + O^* \qquad (5)$$

$$H_2O + O^* \longrightarrow 2OH^{\bullet} \qquad (6)$$

Aldehydes (R.CHO, including formaldehyde H.CHO) can also be photolysed producing hydrogen atoms which eventually result in $OH^{\bullet}$ radicals via reactions already shown in Table 5.1.

Of basic importance to the understanding of polluted urban atmospheres is the photolysis of NO_2 and the subsequent formation of ozone above the background levels. It was noted earlier that only a small proportion of NO_x emissions are in the form of NO_2, the rest being NO. NO emitted into the atmosphere can be slowly oxidized to NO_2 by the reaction with molecular oxygen :

$$O_2 + 2NO \rightarrow 2NO_2 \qquad (7)$$

Photolysis of NO_2 by UV light below 395 nm produces oxygen atoms and subsequently ozone :

$$NO_2 \xrightarrow{UV} NO + O^{\bullet} \qquad (8)$$

$$O_2 + O^{\bullet} \longrightarrow O_3 \qquad (9)$$

However the process is reversed by the reaction

$$O_3 + NO \rightarrow O_2 + NO_2 \qquad (10)$$

so the net result is an ozone level in equilibrium with NO and NO_2 and dependent on the intensity of the solar radiation. The observed levels of ozone are higher than would be predicted on the basis of this limited scheme. High ozone levels imply a high NO_2 NO ratio or rapid NO $\rightarrow$ NO_2 conversion which cannot be achieved by the molecular reaction (7). The type of reactions responsible have already been shown in Table 5.1; they are the transfer of an O-atom from $HO_2^{\bullet}$, CH_3O_2, and other peroxy radicals. Crudely we can say that the photochemical reaction mixture catalyses the NO to NO_2 conversion resulting in the build-up of ozone. Hydrocarbon molecules differ in their ozone production potential. The most important species include the substituted benzenes, such as toluene and the three xylene isomers, and light unsaturated hydrocarbons such as ethene, propene, and but-2-ene. The aromatics are present in high concentration in gasoline and the unsaturated compounds are typical products of engine combustion.

The morning rush hour peak in the NO and hydrocarbon emissions is followed by the gradual conversion to NO_2 and subsequent rise of O_3 which decays as the sun goes down in late afternoon. In an air mass moving downwind from a city, the ozone peak may be worse in the surrounding countryside than in the city centre. During the night most of the reactions that have been described die down but there is an additional route for conversion of NO_2 to nitric acid via the nitrate radical NH_3 which is formed from reaction of NO_2 with O_3. NO_3 which is photolytically unstable in daylight.

The hydrocarbon chemistry in the atmosphere is extremely complex. In addition to reactions with $OH^{\bullet}$ radicals and molecular oxygen similar to those for methane shown in Table 5.1, hydrocarbon species are attacked by the oxygen atoms released in reaction (8) and by ozone. One result of the interaction of the hydrocarbon and NO_x chemistry is the formation of a group of lachrymatory substances including peroxyacetly nitrate (PAN) and peroxybenzoyl nitrate (PBzN). The thresholds for eye irritation for these compounds are only 700 and 5 ppb. respectively. They are formed through the reaction of NO_2 with oxidation products of aldehydes as illustrated in Table 5.2.

Table 5.2. The formation of PAN from aldehydes

$$R\text{—}C(O)\text{—}H + OH^{\bullet} \longrightarrow R\text{—}C(O). + H_2O$$

$$R\text{—}C(O).. + O_2^{\bullet} \longrightarrow R\text{—}C(O)\text{—}O\text{—}O^{\bullet}$$

$$R\text{—}C(O)\text{—}O\text{—}O^{\bullet} + NO \longrightarrow R\text{—}C(O)\text{—}O^{\bullet} + NO_2$$

($R\text{—}C(O)\text{—}O^{\bullet}$ → further reaction)

$$\text{OR} \quad R\text{—}C(O)\text{—}O\text{—}O^{\bullet} + NO_2 \longrightarrow R\text{—}C(O)\text{—}O\text{—}O\text{—}NO_2$$

PAN: R = CH_3; PBzN; R = C_6H_5

PARTICLES

Particle Formation.

The nitric and sulphuric acids formed in the gas reactions described earlier generally undergo further changes. They are both water soluble and will be rapidly absorbed into water droplets if these are present. They may react with solid particles forming sulphates and nitrates. For example, limestone particles (calcium carbonate, $CaCO_3$) can be converted to calcium sulphate, salt particles (NaCl),of marine origin, can be converted to sodium sulphate, Na_2SO_4 or sodium nitrate, $NaNO_3$ with the displacement of hydrogen chloride gas, HCl. However the most common reactions are those involving ammonia :

$$NH_3 + HCl \rightleftharpoons NH_4Cl \quad (11)$$

$$NH_3 + HNO_3 \rightleftharpoons NH_4NO_3 \quad (12)$$

$$NH_3 + H_2SO_4 \rightarrow NH_4HSO_4 \quad (13)$$

$$NH_3 + NH_4HSO_4 \rightarrow (NH_4)_2SO_4 \quad (14)$$

Reactions 11 and 12 are actually reversible and the position of equilibrium depends on the concentrations and the temperature. Significant dissociation of NH_4Cl and NH_4NO_3 occurs during warm summer weather. One end product of the oxidation of SO_2 in the atmosphere is ammonium sulphate, a substance better known as a fertilizer. The degree of neutralization will depend on the supply of ammonia and in the UK sulphate particles are found to be well neutralized on average. The largest source of ammonia is thought to be animal urine.

Particles formed by gaseous reactions are initially very small (<0.1 μm) but grow rapidly either by surface accumulation of material from the gas or by particle-particle coagulation. Once in the size range 0.1 to 2.0 μm, they become relatively stable towards further growth and can remain airborne for periods of days. Most smoke emissions are in this category along with the sulphates and nitrates. At the other end of the size spectrum (2-50 μ m) is coarse dust either emitted from industrial processes or raised by the wind from the ground. The action of road vehicles is another mechanism for raising such dust and sea salt left from the evaporation of spray is also fairly coarse.

Particle Composition

Fig. 5.2 presents a typical breakdowns of the components of the total suspended particulate matter for an urban area based partly on results from a survey in Leeds where the annual average concentration is 40–50 $\mu g\ m^{-3}$. The fine particles are dominated by the ammonium sulphate and nitrates plus carbonaceous material.

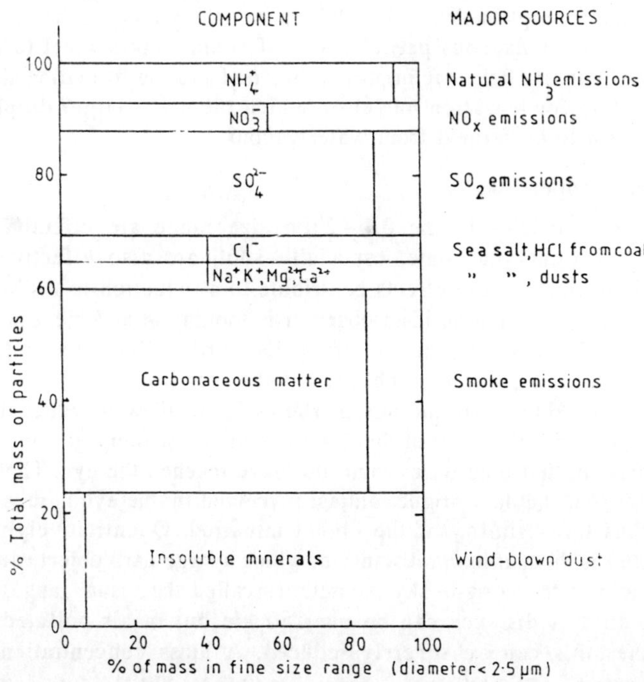

Fig. 5.2. *Composition of atmospheric particles, their origins, and size range. Fine particles* <2.5 µm diameter, course particles 2.5 *µm < diameter* < 15 *µm.*

About one third of this is elemental carbon and the other two thirds organic carbon (*i.e.* high molecular weight solid hydrocarbons) which together comprise 'smoke'. The coarser particles are dominated by wind-blown dusts (clays, silica, limestone, etc.) and include the sea salt component but has smaller amounts of sulphates, nitrates, and carbon with little ammonium.

Deliquescent Behaviour

The particles of water soluble compounds such as sulphates, nitrates, and chlorides will exist in the atmosphere either as solid particles or droplets depending on the relative humidity. For pure compounds the transition from solid to liquid is sharp one. For example, it takes place with salt, NaCl at 75% relative humidity and with ammonium sulphate at 86%.

Particles of mixed composition may continuously grow in size by water absorption from 60-70% up to 100% relative humidity. Insoluble carbonaceous particles are, of course, not subject to this phenomenon. Particles are important in cloud and fog formation since they act as condensation nuclei on which the much larger droplets may begin to be formed from water vapour.

Optical Properties

Fine particles in the 0.1 – 2 μm size range are effective at scattering light and some, especially soot, are also effective at absorbing light. These effects contribute to a reduction in visibility through the atmosphere. If an observer is looking at an object in the distance, he distinguishes it from its surroundings by optical contrast—if it is dark, less light reaches the eyes from the object than from the brighter surroundings. Particles in the air will scatter light from brighter regions out of the line of sight and scatter light into the line of sight that otherwise would not have reached the eye. The net effect is that the dark/bright contrast perceived by the eye is reduced and thus the visibility of the object impaired. Quantitatively, the visibility is the maximum distance at which a large dark object can be seen against the horizon sky (sometimes called the visual range). In clean air this distance can be over 50 km but in air polluted by particles this can be severely reduced. A mass concentration of 200 – 300 μg m^{-3} will reduce the visibility to below 5 km. This situation would be typical of a mid-summer haze on a hot day. It is totally different phenomenon to an early morning mist which is caused by water droplets at a relative humidity near 100%.

DROPLETS

Liquid water occurs in the atmosphere as clouds, mists, and fog within which the concentration of water can be up to 1 g m^{-3}. Smaller amounts of water are also present in association with deliquescent particles as discussed in the previous section. At relative humidities below 95%, this secondary amount will normally be less than 1 mg m^{-3} and will not be considred further.

Water droplets can accumulate pollutants by adsorption of either gases or particles and within the droplets chemical reactions can proceed changing the nature of the adsorbed species. There are several mechanisms for the oxidation of SO_2 sulphuric acid in the aqueous phase. In a cloud situation the most important oxidants appear to be ozone and hydrogen peroxide which is formed from the reaction of two $HO_2^{\bullet}$ radicals. The oxidants must first be adsorbed into water from the gas phase and then result in the reactions :

$$O_3 + SO_2 + H_2O \rightarrow 2H^+ + SO_4^{2-} + O_2 \qquad (15)$$

$$H_2O_2 + SO_2 \rightarrow 2H^+ + SO_4^{2-} \qquad (16)$$

The sulphuric acid formed will be completely dissociated to ions H^+ and SO_4^{2-}. The ozone reactions is inhibited by low pH whereas the H_2O_2 reaction is not. In acidic droplets the oxidation by H_2O_2 is therefore dominant whereas at high pH the O_3 becomes more significant. In a cloud or fog situation where there has been a significant input of particulate pollution the oxidation of SO_2 by atmospheric oxygen catalysed by metal ions (iron Fe^{3+} or manganese Mn^{2+} or by soot can also be important. Although there is partial neutralization by ammonia, the water in polluted fogs and clouds can be much more acidic than in collected rainfall.

It is difficult to distinguish experimentally between the photochemical reaction mechanism leading to H_2SO_4 with subsequent absorption of the acid into water and the aqueous phase oxidation of SC·. However it appears that both routes are important. In the case of nitrogen oxides it is likely that the dominant process is the gas phase oxidation to nitric acid followed by absorption into water. This arises because of the poor solubility of NO_2 and especially NO in water.

DEPOSITION MECHANISMS

Dry Deposition of Gases

At illustrated in Fig. 5.1 the life cycle of an air pollutant normally involves emission, dispersion, and transport, chemical transformation, and finally deposition to the ground. Understanding the rates and mechanisms of deposition is important to the assessment of the environmental impact of many pollutants.

Experimentally we can measure the concentration of the pollutant ($\mu g m^{-3}$) and the total rate of deposition ($\mu g\ m^{-2}\ s^{-1}$). The higher the ground level concentration, the more rapid the deposition but the ratio of these two quantities gives a useful measure of the efficiency of the deposition process. It is called the deposition velocity.

$$\text{Deposition velocity, V} = \frac{\text{deposition rate}}{\text{air concentration}} \text{ (units: m s}^{-1}\text{)}$$

Crudely we can envisage the process of surface adsorption and downward mixing of the SO_2 by turbulent diffusion as a drift of SO_2 to the ground with the calculated deposition velocity. Different surfaces (water, soil, ice, etc.) will have correspondingly different deposition velocities. The description 'dry deposition' is used in all cases even if the removal is to a wet surface.

Deposition to vegetation is rather more complex than deposition to a plane surface. The pollutant's progress may be retarded either by the slowness of diffusion through the air within the canopy of vegetation (*i.e.* between the leaves, etc.) or by the rate of transfer from air to leaf surface (the cuticle) or to the interior of the leaf via stomata. Moisture makes a considerable difference in that transfer to a wet surface is generally faster than to a dry surface. Overall the deposition velocities measured experimentally for SO_2 range from below 5×10^{-3} m s^{-1} to nearly 2×10^{-2} m s^{-1} and a typical value of 1×10^{-2} m s^{-1} is often assumed. Similar considerations apply to other pollutants which are subject to a significant rate of adsorption at the ground or on to vegetation (NO, NO_2, HNO_3, etc.).

Wet Deposition

The term 'wet deposition' is used to describe pollutants brought to ground either by rainfall or by snow. This mechanism can be further sub-divided depending on the point at which the pollutant was absorbed into the water droplets. In-cloud absorption followed by precipitation is termed 'rainout'; below cloud absorption, *i.e.* pollutants collected as the raindrops fall, is termed 'washout'. The rate of removal of a pollutant by washout will increase in proportion to the rainfall rate. For SO_2 about half of the gas below the clouds will be removed in two hours of heavy rain. The rates for the nitrogen oxides are lower due to their reduced solubility in water. In remote areas the majority of the wet deposition of sulphur appears to be due to rainout. Washout becomes relatively more significant near the sources of pollution where the gas concentrations are high.

Dry deposition and wet deposition are of comparable importance in the case of SO_2. Dry deposition is most significant where ground level concentrations are high, in other words close to the sources.

The height of the emission makes only a very small difference (perhaps 10%) to the amount of long-range transport. This arises because a slightly higher proportion of low level emissions are lost by dry deposition to the ground near the sources than in the case of high level emission.

Another mechanism of deposition is when fog or cloud droplets are removed directly to the ground or to vegetation. This is termed 'occult deposition'. It becomes significant at elevated locations such as mountains or hill tops. There is the potential for more severe damage to foliage than with acid rain since, as was previously mentioned, polluted fog or cloud droplets can contain much higher concentrations of acidic pollutants than raindrops.

Deposition Particles

The mechanism of deposition of particles depends on the particle size. Large particle with diameter greater than 10 μm fall slowly by gravitational settlement. The larger the particles, the more rapidly they fall. The sedimentation velocities for particles of density 2 g cm^{-3} are as follows :

Diameter μm	Velocity m s^{-1}	Diameter μm	Velocity m s^{-1}
5	1.5×10^{-3}	50	1.4×10^{-1}
10	6.1×10^{-3}	100	4.6×10^{-1}
20	2.4×10^{-2}	150	8.0×10^{-1}

Particles larger than 150 μm diameter, falling at over 1 m s^{-1} remain airborne for such a short time that they do not need to concern us as air pollutants. Particles less than 5 μm have sedimentation velocities which are so low that their movement is determined by the natural turbulence of the air, just as for gases.

Intermediate particles, between 1 and 10 μm diameter, can be removed by impaction onto leaves and other obstacles. Particles in the 0.1 to 1 μm range which include most of the nitrates and sulphates, are only removed very slowly by dry deposition. The deposition velocities are of the order of 10^{-3} m s^{-1} a factor of 10 lower than for SO_2. The most likely route for their removal is rain-out following water vapour condensation and droplet growth in clouds. Wash-out is not very efficient for these fine particles although it becomes more significant for larger particles such as course dust.

ACID RAIN

The Effects

The phrase 'acid rain' has come to be used very loosely to mean almost anything to do with air pollution whether or not rain is actually involved. Three particular effects have received most attention. Historically the first of these was the increased acidification of lakes and streams in Scandinavia leading to loss of fish and other aquatic organisms. This was attributed to the acidity of rain polluted by sulphur and nitrogen oxides and the UK was blamed as being the chief culprit.

The second type of environmental damage is the die-back of forests of Central Europe. The effects became very noticeable in Germany in the early 1980s and are worst in the south-west (the Black Forest) and on the eastern border with Czechoslovakia, a country which shares the same problem. Since then other countries have reported similar phenomena. Although the reasons for the damage are still being debated, it seems likely to be a combination of factors. Predisposing factors include drought and high altitude. Some forests in the Eastern European countries suffer from the effect of very high SO_2 levels but this is not the case further west. Possible mechanisms include the effect of ozone initiating the attack and subsequent further deterioration being due to acid rain or acid mists and fogs. Acidification of the ground with consequent effects on the soil chemistry may also contribute. If we accept that both ozone and acidic precipitation are important for forest damage then attention must be given not only to SO_2 emission but also to NO_x and hydrocarbon emissions as the precursors of ozone.

Surveys indicate that forest damage exists but the importance of pollution among the various possible factors is not well established. It is clear that the ozone levels are lower than in Central Europe and the occasional high values tend not to persist due to our maritime climate. Few forests exist at the high elevations common in Germany.Direct comparison between the two countries is therefore difficult.

The third problem is the attack on stonework and the decay of many of the most famous buildings constructed of limestone. Both wet and dry deposition of sulphur dioxide are involved. Under moist conditions SO_2 or sulphuric acid will convert calcium carbonate to gypsum, $CaSO_4$. Since the sulphate is more soluble than the carbonate, the reacted stone can be removed by dissolution. The solid gypsum also occupies a larger volume than the original carbonate and this leads to spalling of material from the surface. A combination of these factors leads to a rate of loss of stone which depends on the deposition rate of SO_2 to the surface. Deposition to a moist surface is more rapid than to a dry surface so the fraction of time the surface is wetted as well as the SO_2 concentration is important. It is generally assumed that in urban areas the dry deposition of SO_2 gas is the major factor. Control of urban SO_2 level requires a completely different approach to the control of national emissions, which is the objective of the wider European movement for the reduction of acid rain. Reduction is sulphur content of fuel oils and further conversion from coal and solid fuels to natural gas or electricity are all relevant. Limitations on power station emissions may help in some areas but in other areas will make no difference.

Rainwater Composition

Even in the absence of air pollutants, rain-water is slightly acidic (pH 5.6) due to atmospheric carbon dioxide. 'Acid rain' therefore refers to rain with a pH below about 5.

A typical ionic composition in units of micro-equivalents per litre of a fairly polluted rain-water with high ionic concentrations and a pH of 4.1 might be as follows :

Cations H^+ 80, NH_4^+ 120, Na^+ 80, K^+ 10, Ca^{2+} 60, Mg^{2+} 25

Anions SO_4^{2-} 170, NO_3^- 90, Cl^- 120

Qualitatively we can see that the acidity originates from SO_2 and NO_x which have been oxidized to H_2SO_4 and HNO_3. Overall about two-thirds of the acid is associated with SO_2 emissions and one-third with NO_x emissions. The acid has been partially neutralized by ammonia and other ions such as Ca^{2+} which may have originated as calcium carbonate, $CaCO_3$. In the ground NH_4^+ can release H^+ during the process of being oxidized to NO_3^- by bacteria. The H^+ can then be leached out into steams. From the point of view of freshwater acidification, ammonia should therefore not necessarily be regarded as an ally even though it reduces the rain-water acidity.

Most of the sodium, chlorine, and magnesium in precipitation originate from set salt. Non-marine chloride levels are only significant near major sources such as coal-fired power stations.

CHAPTER 6

Water : Resources And Pollution

THE NATURE AND COMPOSITION OF WATER

Natural Characteristics

Water, H_2O, hydrogen oxide, is an extraordinary chemical compound of fundamental environmental importance. The electronic

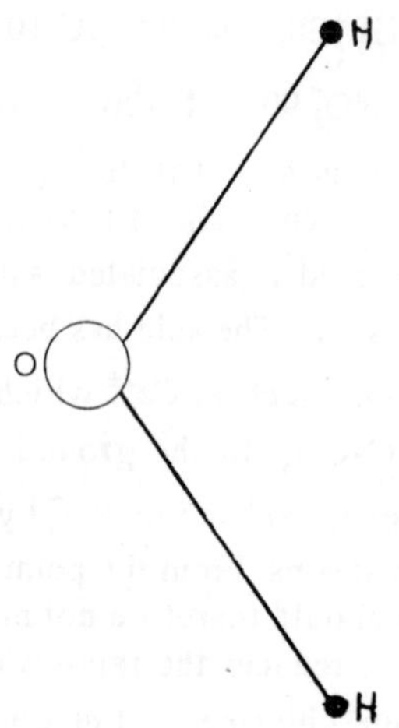

Fig. 6.1. *Diagram of the water molecule*

structure of its molecules impart the very special chemical and physical properties which lead to water being described as 'the universal solvent' or 'the liquid of life'. Water in the gaseous state is made up of single molecules, the structure of which is usually represented as shown in Fig. 6.1. Water in the liquid state is made up of groups of molecules associated together by linkages (hydrogen bonding) between each of the hydrogen atoms of one water molecule and the oxygen atom of an adjacent water molecule, as depicted in Fig. 6.2. In ice the water molecules are associated in tetrahedral structure formed by four molecules arranged around a central molecule. It is the nature of the molecular structure and the ways in which these molecules associate in hydrogen bonding which yield, among other chemical and physical characteristics, the general solvent power, the chemical reactant power in biological processes, and the heat storage and transfer power of water as a key environmental resource.

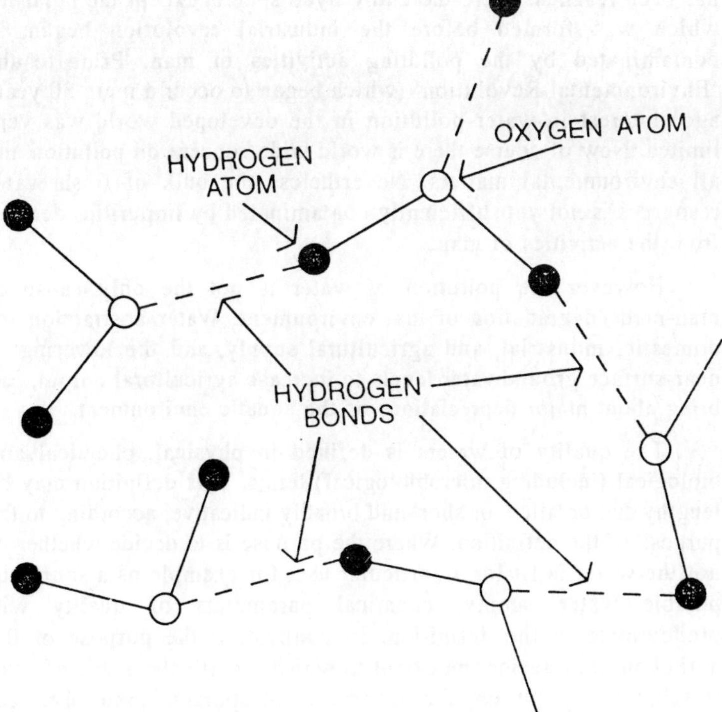

Fig. 6.2. *Hydrogen bonding in water*

The Earth's water resources, the 'hydrosphere', consists of the oceans and seas, the ice and snow of the polar regions and mountain glaciers, the water contained in surface soils and underground strata, and the water in lakes, rivers, and streams. Less than 1% of these resources consists of freshwater, some 2% is freshwater ice located mainly in the polar regions, and the remaining 97% or so consists of seawater and sea ice. The annual evaporation of water from the hydrosphere, and its return as rainfall (the hydrological cycle) amounts to about $260 \times 10^{12}\,m^3$. The total water content of the atmosphere is about $7 \times 10^{12}\,m^3$, indicating that atmospheric water is replaced on the average some 37 times a year.

For all practical purposes it can be said that the waters of the hydrosphere were of natural quality until the Industrial Revolution in Europe and North America initiated the development of technology driven by the energy of the fossil fuels coal and oil. Now the stage has been reached where the entire hydrosphere, except the polar ice which was formed before the industrial revolution began, is contaminated by the polluting activities of man. Prior to the 'Environmental Revolution' (which began to occur a mere 30 years ago) interest in water pollution in the developed world was very limited. Now of course there is world-wide concern on pollution and all environmental matters. Nevertheless the bulk of fresh water resources is not yet sufficiently contaminated by impurities derived from the activities of man.

However the pollution of water is not the only cause of man-made degradation of that environment. Water abstraction for domestic, industrial, and agricultural supply, and the lowering of near-surface groundwater levels to increase agricultural output, can bring about major depreciations of the aquatic environment.

The quality of waters is defined in physical, chemical and biological (including microbiological) terms. This definition may be lengthy and detailed, or short and broadly indicative, according to the purpose of the definition. Where the purpose is to decide whether or not the water is fit for a particular use, for example as a source of potable water supply, chemical parameters of quality will predominate in the definition. In contrast, if the purpose of the definition is to decide the extent to which a particular reach of river is subject to pollution, the occurrence of sporadic toxic chemical pollution is more likely to be deduced from observation of the biology and the physical characteristics of the river regime than by other means. Of course the proof of occurrence of any such sporadic chemical pollution will inevitably be expressed in chemical terms. Furthermore the parameters of water quality which are specified as limitations of the acceptability of a water for particular use purposes, or as a general environmental resource, are predominantly chemical and biochemical ones. Such specifications usually appear as 'quality standards', 'quality criteria', or 'quality guidelines' promulgated by international bodies and individual nations.

Rainfall, even when it was in its wholly natural state, was quite impure water in the scientific sense. Its physical characteristics in the liquid phase are that it would normally be naturally clear, colourless, and odourless with a temperature varying from ambient to colder depending on the nature and height of its precipitation. Rainfall naturally contains materials dissolved from the atmosphere. Firstly it contains the atmospheric gases—mainly nitrogen, oxygen, and carbon dioxide dissolved from the atmosphere. Secondly it contains matter dissolved from impurities in the air derived from the earth's surface—such as sea spray, volcanic emissions, wind-borne dusts, hydrogen sulphide and methane from anaerobic decomposition, and volatile organic compounds derived from land and aquatic plants. Thirdly it contains traces of ozone and other gases derived from chemical reactions, triggered by solar and cosmic radiation, of naturally occurring materials in the atmosphere. While it is essential to understand this natural chemistry of rainfall, it is the chemical consequences of the man-made pollutions of the atmosphere, which cause the greatest concern. Thus, apart from the effects of atmospheric pollution, the natural quality of rainfall is determined mainly by seawater and dust take-up, varying according to weather, locality, and major volcanic events. Rainfall naturally contains around 10 to 20 mg L^{-1} of dissolved solids according to circumstances. It is naturally somewhat acidic in character, with a pH value of around 5.5, because of its solution of atmospheric acid gases, mainly carbon dioxide.

On reaching the ground, rainfall picks up more impurities naturally from vegetation and the ground surface. If the ground is permeable, all or some of the rainfall will pass underground according to circumstances. Some will be evaporated and the balance will run off to streams and rivers and via underground strata *en route* to the sea. On the average some 70% of the rainfall evaporates directly, or indirectly via plant transpiration. Of the remaining 30%, more travels through underground strata than along rivers and streams, and about 90% of the total run-off reaches the oceans. Natural run-off from bare, substantially impermeable, hard rock gathers material from gradual weathering of the rock to yield low concentrations of dissolved calcium, Magnesium, sodium, and potassium. Traces of other materials will be present according to the chemistry of the rock. In contrast, where the rock underlies a cover of peat and associated vegetation the take up of stable organic matter will be high. The run-off from peat areas is usually acid in the p^H range 5 to 6, and has a pronounced yellow-brown colour, highly so in times of heavy rainfall. A similar situation applies to the run-off from woodland areas where ground cover of leaf mould produces a similar effect.

Table 6.1. Analysis of run-off from a boulder clay catchment

	$mg\ l^{-1}$ (*unless otherwise stated*)
Ammonia nitrogen	0.2
Chemical Oxygen Demand (dichromate value)	4.0
pH Value	7.8 units
Chloride (Cl)	50
Nitrate (N)	6.0
Total hardness ($CaCO_3$)	410
Non-carbonate hardness ($CaCO_3$)	250
Magnesium (Mg)	8.0
Phosphate (PO_4)	0.6
Sulfate (SO_4)	140
Silicate (SiO_2)	6.0
Iron (Fe)	0.05

The run-off from clay soil is characterized by its content of calcium sulfate and calcium carbonate, compounds which are naturally abundant in clays. Turbidity in the water is often evident because of the presence of colloidal clay particles. An analysis of a typical water running off a clay area in South-east England is shown in Table 6.1. The run-off from clay soils during prolonged heavy rainfall, particularly from ground with little plant cover, is heavily loaded with suspended particles of clay and fine sand derived from the clay. While during heavy rainfall the water referred to in Table 6.1 would contain up to 0.1% suspended soil particles, the larger rivers of the world subject to tropical rainfall intensity carry relatively immense concentrations of solids in suspension.

Freshwater Rivers, Streams, and Lakes

By and large the natural quality of rivers and streams (henceforth collectively referred to as rivers) at any point reflects the quality of the upstream contributions of surface run-off and groundwater discharge. Similarly, but to a lesser degree according to circumstance, the natural quality of a lake reflects the quality of the inflows of water that maintain the lake level. However these waters, open to the energy of sunlight, the solution of oxygen from the air, and containing the mineral nutrients sufficient to supports plant growth, naturally become the media for the growth of aquatic biota. These biota can be divided into four categories, namely :

the phytoplankton, being plants mainly algae, floating within the water ;

the zooplankton, which include bacteria and protozoa, floating within the water ;

the rooted plants, mostly growing from the river or lake bed, but some floating on the water ;

the large animals which are either free swimming or are attached to the bed or the larger plants.

The nature, variety, and abundance of these biota—the biology of the water—is essentially determined by the physics and chemistry of the water. However there is a feed-back loop in this system which results in the growth of the biota exerting an impact on the physics and chemistry of the water—for example the effects of phytoplankton growth in increasing the oxygenation and alkalinity of the water, and in transferring some of the calcium, carbonate, and phosphate content of the water to the river or lake bed.

The natural shape of the channel of a river at any point is determined by the rate of flow (discharge) of water and the topology and geology of the terrain through which the river passes. These factors determine the gradient of the river, its velocity of flow, its depth and width, its shoals and pools, and the nature of the river bed at various points—silt, gravel, or bare rock as the case may be. Even as the growth of river biota has a secondary effect on the physical and chemical characteristics of the water, so the growth of the biota can influence the effective shape of the river channel. A good example of this is the heavy growth of rooted water plants which occurs in the shallow, calcareous waters of chalk streams. Heavy growths of these plants, if not controlled by judicious cutting and removal from the rivers, often so reduce the flow cross-section as to cause flooding during heavy summer storms. Another major cause of quality variability is the biochemical activity which proceeds within micro-organisms in the water, particularly bacteria, as key factors in the operation of the well known carbon and nitrogen cycles and other biochemical transformations. The most important of these are summarized in Table 6.2. It needs to be borne in mind of course that straight-forward chemical reactions occur also—for example the precipitation of metals from solution in the water.

Table 6.2. Main biochemical processes proceeding in water

Process	*Requirements*[a]	*Main end products*
Oxidation of organic carbon	Dissolved oxygen, $T > 0°C$	CO_2, H_2O
Oxidation of ammonia	Dissolved oxygen, $T > 4°C$	NO_3^-, little NO_2^-
Reduction of nitrate	Dissolved oxygen absent, $T > 0°C$	N_2, some N_2O
Reduction of organic carbon	Dissolved oxygen absent nitrate absent, $T > 4°C$	CH_4, CO_2

WATER POLLUTION

It is part of the natural scheme of things that man, a saprophytic animal, should cause environmental pollution in almost all he does. Fortunately it is also part of the natural scheme of things that man, blessed with thinking ability, should recognize the need to control pollution and devise technological and administrative means of effecting this control. That this recognition is now apparently widespread does not necessarily mean that the required controlling action will be Man-made pollution of water is divided into two kinds, namely into point sources and non-point sources. The most important of these are listed in Table 6.3. As might be expected, the point sources are mainly discharges of wastewaters from sewage works, factories, and farms, while the non-point sources arise mainly from specific categories of general land use (*e.g.* high intensity farming). As regards the type of pollution which occur, these can be listed according to the effect exerted by the polluting matter, as follows :

Table 6.3. Main point and non-point sources of pollution

Point sources,	*Non-point sources*
Discharges from sewage treatment works to rivers	Run-off and underdrainage from agricultural land into rivers
Discharges of industrial wastewaters to rivers	General contamination of recharge rainfall to outcropping aquifers
Discharges of farm effluents to rivers	Septic tank soakaways into permeable strata
Discharges from small domestic sewage treatment plants to rivers	Wash-off of litter, dust and dry fallout, from urban roads to rivers
Discharges by means of well or borehole into underground strata	General entry of sporadic and widespread losses of contaminants to rivers
Discharges of collected landfill leachate to rivers	Seepage of landfill leachate to underground strata and to rivers

(a) Substances acutely toxic to man and/or aquatic flora and fauna (*e.g.* lead, mercury, cadmium, cyanide, pesticides) ;

(b) Substances which are hazardous to man and/or to flora and fauna in causing chronic, or long dormant cumulative, harm (*e.g.* polynuclear aromatic hydrocarbons, chlorophenols, trihalomethanes) ;

(c) Substances at very low concentrations which are not highly toxic but which can either be rendered acutely toxic by biochemical transformation in the water (*e.g.* methylation of inorganic mercury) or by bioconcentration (*e.g.* triphenyl tin, see Case Study 1) ;

(d) Substances which add to the load of biochemical oxygen demand in the water or in benthal (bottom) deposits, (*e.g.* sewage effluent, food-industry wastewater, farm wastes) ;

(e) Substances which add to the eutrophication (plant-nutrient content) of the water (*e.g.* sewage, farm wastes) ;

(f) Substances which have a detrimental effect on the physical appearance of the water (*e.g.* oil, detergent foam, litter, suspended matter) ;

(g) Substances which only have a polluting effect at relatively high concentrations in water (*e.g.* mineral salts such as sodium chloride) ;

(h) The waterborne organisms which are pathogenic to man (*e.g.* *Salmonella*, *Cholera vibrio*).

In this list, the organic substances in categories (a), (b), and (c) exert their polluting effects at very low ($\mu g\ L^{-1}$) concentrations. The inorganic substances exert their effects at higher concentrations (in the range of $mg\ L^{-1}$). Those in categories (d) to (g) exert their effects at multi $mg\ L^{-1}$ concentrations. The hygienic safety of waters, through which man may be exposed to infection by the organisms in category (h), can only be high where bacteria of faecal origin cannot be detected in 100 ml of water (per litre in the case of a plaque forming unit of polio virus).

DIFFUSE POLLUTION

Pollution from non-point sources, or diffuse pollution, in its very nature presents difficult control problems. If it is known that a river, lake, or an aquifer is polluted by a particular substance or group of substances, but the point of access of this pollution to the water cannot be located and no particular person or corporate body can be proved to have caused or knowingly permitted the pollution to occur, the normal processes of penalizing polluters and making them pay for remedial and/or preventive measures are thwarted. Special research and administrative action has then to be taken, and since this is a difficult and slow process, such action can be almost guaranteed to take place too late to avoid major damage arising. There are three main problems before us. The first is the acidification of waters caused by acid rain and the second is the high and still rising level of nitrate in many waters, particularly groundwaters. The third relates to water contamination by leachate from landfill sites and spills of chemicals.

Acidification of Waters

The acidification of rivers and lakes by acid deposition from the atmosphere is essentially a local problem. Much of the acidity is a part of the consequences of the major problem of atmospheric pollution in industrial areas.

The effects of acid deposition varies greatly according to the type of soil on which it falls. Alkaline soils based on limestone can neutralize large amounts of acid whereas soils based on peat or granite cannot do so. There are two practicable ways in which the impact of acidification can be cased. One is to reduce the emission of the atmospheric pollutants. The second is to add a neutralizing alkali (such as powdered limestone) to the acid-sensitive areas. The key link between the emissions and their ecological impact is the transfer of the acidity from deposition to run-off to rivers and lakes, and understanding of this link is essentially a matter of the chemistry of the interactions between the soil and the soil water. In essence the lightly buffered soils, thin and base poor, are the most sensitive to acid deposition and are the main generators of acid rivers and lakes from that deposition.

Nitrate in Waters

The EC directive on the quality of water intended for human consumption sets the maximum acceptable limit for nitrate in water at 50 mg L^{-1} (11.3 mg L^{-1} nitrate nitrogen).

Nitrate problems in groundwaters arise mainly because of heavy loadings of nitrogenous matter on land through which percolating rainfall recharges the underlying aquifer. The only practicable way of reducing the rate of increase of nitrate concentration, and eventually reducing the actual concentration in the groundwater, is to reduce the intensity of agricultural activity which produces the heavy nitrogenous loading. There is plenty of evidence that nitrate levels in many groundwaters have increased greatly over recent decades, in some case to the extent that water abstractions from boreholes has had to be curtailed or even abandoned. Further, computer models indicate that this position will worsen unless remedial action is taken.

Water Pollution via the Contamination of Land

The subject of land contamination by the disposal of solid wastes and sludges, and spillages of polluting liquids, to land will be dealt with in Chapter. The sound management of water resources demands of course that the entry of foul leachates from landfill sites and other sources to waters should be kept under effective control. The greatest difficulty arises when valuable groundwater sources are rendered unfit for use, often by quite small levels of contamination, from leachates from landfill or spillages. It is usually most difficult, and most costly, to decontaminate the water source within a reasonable time, and there are many groundwater sources that have been abandoned because of the virtual irreversibility of the contamination. The modern techniques of landfill control and management are developing such that new leachate pollution can be avoided and serious existing contamination greatly reduced.

WATER PURIFICATION FOR SUPPLY AND WASTEWATER TREATMENT

Water Purification for Supply

The primary requirement regarding the quality of public water supplies is the public health requirement that the water should be 'wholesome' for drinking. Wholesome in this context is generally interpreted to mean, as stated in dictionaries, 'promoting or conducive to health'. Over the past 30 years or so, following the lead of the World Health Organization, most of the countries of the world have adopted standards which prescribe the quality criteria with which water supplies should conform to be considered satisfactory for drinking. Some of these countries have simply adopted the WHO's recommendations. Others have adopted most of the WHO's recommendations but have made variations to a few of the recommendations. The USA has adopted standards which are largely in line with the WHO criteria but are more broad ranging and detailed, and of course the European Community Directive on the Quality of Water Intended for Human Consumption came into effect in 1985. Given that a public water supply complies with requirements for drinking water, it can be taken as axiomatic that the water is suitable for agricultural purposes. However public water supplies are not sufficiently pure for use for many industrial purposes—for example in steam raising, in the chemical and pharmaceutical industries, and in the health care and electronics industries. Each industry is expected to take its own steps to produce, from the available public supply, the precise level of purity of water it requires for its particular purposes.

While the quality of public supply, like any other water, is described in physical, chemical, and biological terms, it is the microbiological quality which matters the most—simply because it is very easy for water to be contaminated by the micro-organisms causing human disease and major epidemics. The physical characteristics of public supply are measured in terms of its colour, turbidity, odour, taste, and pH value. The chemical characteristics cover a wide and complex field, embracing low concentrations of the acutely toxic inorganic chemicals, very low concentrations of the hazardous organic micro-pollutants such as polycyclic aromatic hydrocarbons, haloforms as produced in the chlorination of organic residuals in water, pesticides, mercury, and cadmium. Also specific inorganic chemicals such as fluoride, nitrate, nitrite, magnesium, and sulphate which at particular levels of concentration are or may be harmful to health are included. Other substances such as chloride, iron, manganese, and residuals of organic matter of vegetable or animal origin, which in one way or another could affect the acceptability of a water for drinking, are included. Table 6.4 sets out the normal chemical characteristics of public water supplies drawn from different sources, and Table 6.5 summarizes the more important chemical and microbiological characteristics of public water supply which feature in the quality standards of the EC directive on drinking water.

The importance of the microbiology of public supply concentrates on the organisms responsible for waterborne human disease. These may be classified as bacteria, viruses, and other organisms such as worms, flukes, and protozoa which are known to be associated with the waterborne spread of disease. Table 6.6 gives a very brief summary of the main aspects of this microbiology. In the day to day control of the microbiology of public water supply, since it is not practicable to examine waters frequently for the presence of all the organisms that cause disease, standard bacteriological examination of water in a relatively simple but effective way is carried out. This involves assessing the numbers of bacteria

Table 6.4. Normal chemical characteristics of some public water supplies (ppm)

	Source of water			
	Chalk borehole (softened)	Upland river	*Lowland river*	*Lowland reservoir*
pH value	7.3	7.3	7.9	7.6
Colour (units)	0	2	5	4
NH_3 (N)	0	0.10	0.02	0.03
NO_3(N)	3.0	1.2	4.0	1.7
Total solids	220	80	410	430
Total hardness ($CaCO_3$)	150	45	260	225
Non-CO_3 hardness	10	25	135	70
Alkalinity ($CaCO_3$)	110	20	130	160
Chloride	16	15	35	55
Iron	0.01	0	0.02	0
Manganese	< 0.01	<0.01	<0.01	<0.01

Table 6.5. Some aspects of water quality specified in the EC drinking water Directive

Contaminant	*Maximum admissible concentration*
Acidity/alkalinity	pH range 5.5–9.5
Colour	20 units
Turbidity	4 units
Iron	200 $\mu\ l^{-1}$
Manganese	50 $\mu g\ l^{-1}$
Aluminium	200 $\mu g\ l^{-1}$
Nitrate nitrogen	11.5 $mg\ l^{-1}$
Lead	50 $\mu g\ l^{-1}$
Pesticides (individual substances)	0.1 $\mu g\ l^{-1}$
Pesticides (total)	0.5 $\mu g\ l^{-1}$
E. coli	Not detectable in 100 ml of water
Coliform bacteria	95% of samples should be coliform free

Escherichia coli (*E.coli*) and of the coliform group present. The former is indicative of the faecal pollution of water while the presence of the other members of the group is indicative of pollution from animal

Table 6.6. List of the organisms mainly responsible for waterborne disease (temperate climate),

Type of organism	*Name*	*Source*	Disease
Bacterium	*Salmonella typhi*	Man	Typhoid fever
Bacterium	*Salmonella typhi*	Man	Paratyphoid fever
Bacterium	*Vibrio cholerae*	Man	Cholera
Bacterium	*Shigellas*	Man	Bacterial dysentry
Bacterium	*Salmonella* group	Man and animals	Gastro-enteritis (food poisoning)
Bacterium	*Leptospira interohaemorrhagiae*	Rats	Weil's disease
Protozoa	*Cryptosporidium parvum*	Man and animals	Diarrhoeal illness
Amoeba	*Entamoeba hystolica*	Man	Amoebic dysentry
Tapeworm	*Taenia saginata*	Man via cattle	Beef tapeworm

sources, but not necessarily of faecal pollution. The EC directive and the WHO guidelines both require *E.coli* to be absent from public supply, whereas a little latitude, under prescribed circumstances, is permitted as regards the presence of coliform group organisms in water in distribution mains. The macrobiology of public water supply is important in relation to the efficiency of water purification and to the aesthetic quality of the water. Concern here relates to the presence of algae and invertebrate animals in the water, on which further information is given to Table 6.7.

Table 6.7. Algae and invertebrate animals of concern in public water supply

Type of organism	*Common species or genera*	Cause of presence in public supply
Small single-celled Chlorophyta (green algae)	*Chlamydomonas* *Arkistrodesmus* *Scenedesmus*	*Passing through waterworks filters*
Bacillariophyta (diatoms)	*Stephanodiscus*	Passing through waterworks filters
Cyanophyta (blue-green algae)	Various species	Passing through waterworks filters
Cyanophyta	Various species	Decay of algae causes taste and odour in water
Xanthophyta (yellow-greenalgae)	*Chryptomonas* *Rhodomonas*	Passing through waterworks filters
Xanthophyta	*Synura* *Peridinium*	Causes cucumber taste Causes fishy odour
Worms	*Nias*	Can pass filters and infest water mains
Rotifera	Various species	Passing through filters
Fly larvae	*Chironomid* sp.	Can pass filters and infest water mains
Crustacea	*Daphnia, Cyclops, Asellus*	Can pass filters and infest water mains

The nature and extent of the treatment to be given to any particular raw water, to produce a public supply meeting specified quality standards at minimum cost, depends, entirely on the nature and quality of the raw water. The minimum treatment, of disinfection using chlorine, would be appropriate for a deep groundwater source in a rural area, or for water from an upland impoundment not open to public access. At the opposite extreme, water drawn from the lowland reaches of a river, containing a high proportion of sewage or industrial wastewater discharge upstream, would need extensive and very carefully monitored treatment. The conventional techniques for the treatment of raw waters for public supply are set out in Table 6.8., and Fig. 6.3 shows the layout of a typical waterworks treatment plant.

Wastewaters and Wastewater Treatment

The present system of wastewater disposal worldwide is based on the provision and operation of public sewerage reticulations and sewage treatment plants in urban areas, and the private provision of drains and wastewater treatment arrangements at industrial premises and stock rearing farms. Much of the industrial wastewater produced in urban areas is disposed of, with or without pre-treatment, into public sewers with the consent of the water services. This general position applies in many countries but in others, particularly developing ones industrial wastewater disposal direct to rivers is the norm. Wastewaters produced on farms are usually too strong in the load of organic matter they carry for disposal to public sewerage systems. Instead, efforts are made to contain the farm wastes on the producing farm where the fertilizer value of the wastes can be realized.

Overall world-wide, most of the wastewater produced in urban areas in finally discharged to rivers of tidal waters. One of the major urban water pollution problems is that of disposal of storm sewage. This arises particularly in the older areas of developed countries where the sewerage systems are laid wholly or in part as 'combined sewers' —that is they receive both foul water (sewage and industrial wastewater) and rainfall run-off from roofs, and roads and other paved areas. In times of heavy rainfall these combined sewers become surcharged and to prevent serious and foul flooding the sewers have to be relieved of some of the overload of diluted sewage they carry. This is done via devices known as storm sewage overflows which discharge to convenient watercourses. These are obnoxious ways of managing urban drainage and the only reasonable long term remedy is to steadily renew the old sewers on a separate system of foul sewers and surface water sewers.

Table 6.8. Summary of treatment processes used in purification of public water supply

Process	*Purpose*
Raw water storage (short term)	Sedimentation. Die-off of faecal organisms. Balancing of intake water quality. Raw water reserve
Raw water storage (long term)	Oxidation of organic matter. Partial removal of NO_3, HCO_3, PO_4, SiO_2, by algal uptake
Chemical precipitation using $Al_2(SO_4)_3$ or activated SiO_2, or Fe salts plus polyelectrolytes	Coagulation, flocculation, and settlement of turbidity and colour
Microstraining	Straining through very fine-mesh rotary screens
Rapid filtration	Rapid up-or down-flow filtration through sand
Slow sand-filtration	Filtration plus bio-oxidation by slow gravity downward flow
Chlorination and/or ozonation or UV	Disinfection
Softening by lime, lime-soda, or iron exchange	Removal of Ca and Mg hardness (no longer fashionable)
Activated carbon treatment by powder addition before filtration or passage through active carbon filters	Reduction in residual organic matter
Desalination by flash distillation or reverse osmosis	Production of freshwater from saltwater or super-purification of wastewater

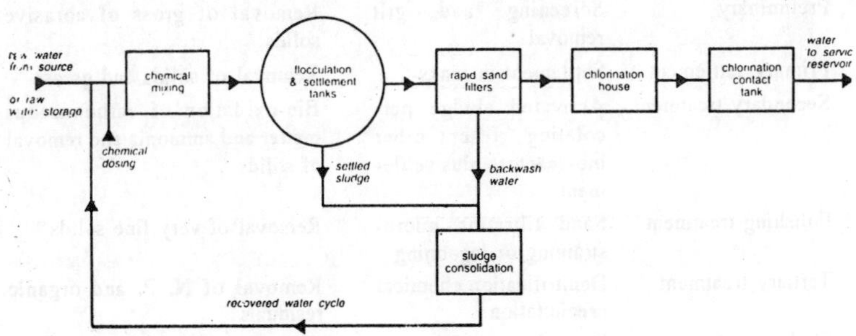

Fig. 6.3. *Layout of a typical water treatment works for public supply*

Sewage Treatment

Modern sewage treatment employs three basic processes, namely :

(a) the removal of polluting matter from the sewage flow as solids, or slurries of solids in water (sludges) ;

(b) the removal of polluting matter from the sewage flow and from separated sludges by accelerated natural processes of biochemical breakdown ;

(c) the separation of water from sludges to reduce the volume of sludge for disposal.

Table 6.9 sets out the conventional descriptions of these processes and the type and nature of the treatment given, and Fig. 6.4 shows the layout of typical sewage treatment works.

Primary settlement should reduce the suspended solids (dry) content of the sewage from about 300 ppm to about 120 ppm, and the BOD from about 250 ppm to about 120 ppm. Secondary treatment processes are based on placing the degradable carbonaceous and nitrogenous matter, carried in the settled sewage, into intimate contact with oxygen and a sufficient surface of biomass for an appropriate period of time. Where the reactor containing the process is a percolating filter the retention time is about 20 minutes, but the structure of the interior of the filter provides a large void space, a large reactive bio-surface,

Table 6.9. Summary of main sewage treatment processes

Process	*Type of treatment*	*Basic purpose*
Preliminary	Screening and grit removal	Removal of gross of abrasive solids
Primary settlement	Settlement in tanks	Removal of solids and grease
Secondary treatment	Activated sludge percolating filter other bio-reactors plus settlement	Bio-oxidation of carbonaceous matter and ammonia and removal of solids
Polishing treatment	Sand filtration micro-straining or lagooning	Removal of very fine solids
Tertiary treatment	Denitrification chemical precipitation	Removal of N, P, and organic residuals
Sludge treatment	Digestion thickening dewatering drying	CH_4 production. Preparation for disposal

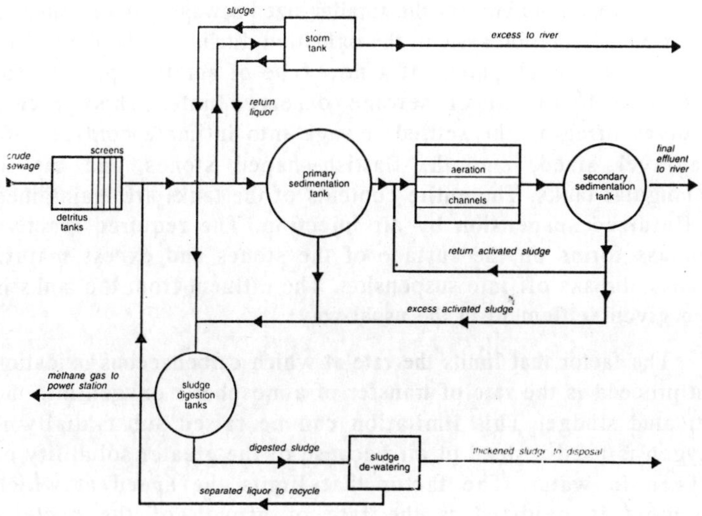

Fig. 6.4. *Layout of a typical sewage treatment works using the activated sludge process*

natural ventilation through underdrains, a low volumetric loading of flow to capacity, and a large population of grazing fauna. Where the reactor is an activated sludge aeration tank (channel, square, or circular) the retention time is 4 to 6 hours. The air requirement, supplied as blown diffused or coarse bubble air, or mechanical surface aeration must be sufficient to maintain a residual of dissolved oxygen. A mixed liquor suspended solids concentration (activated sludge) of around 2000 to 4000 ppm is the normal range. The latter figure applies where good nitrification (conversion of reduced nitrogen to nitrate) is required and is normally accompanied by the re-aeration of the activated sludge being returned to the flow of settled sewage at the inlet to the aeration tanks. The effluent from secondary treatment is passed to settlement tanks which are similar to, but about one-third of the capacity of, primary settlement tanks. Domestic sewage that has been given full primary and secondary treatment, including 6 hours aeration in a diffused air activated sludge plant should normally contain less than 15 ppm suspended solids, about the same level of 5 day BOD and an ammonia nitrogen content of less than 10 ppm.

Several major modifications of the activated sludge process have been developed, mainly for the smaller sized sewage works—such as the extended aeration system, the oxidation ditch, and the deep-shaft system. The development of a new type of aeration process for secondary treatment of sewage deserves note. This process involves bringing the settled sewage into intimate contact with peagravel sized, smooth, flattish-shaped stones and air in rectangular tanks. The entire contents of the tanks are maintained in fluidized suspension by air injection. The required reactive biomass forms on the surface of the stones and excess mature biomass breaks off into suspension. The effluent from the tanks is then given settlement in the usual way.

The factor that limits the rate at which carbonaceous oxidation can proceed is the rate of transfer of atmospheric oxygen into the activated sludge. This limitation can be raised substantially if oxygen is used instead of air because of the greater solubility of oxygen in water. The factor that limits the speed at which ammonia is oxidized is the rate of growth of the bacteria responsible for oxidizing ammonia to nitrate. These have a doubling time of about 1 day, whereas the bacteria responsible for the oxidation of carbonaceous matter have a doubling time of 20 minutes.

Domestic sewage sludge contains about 15% saponifiable fat, 10% fibre, 10% protein, 35% ash, 0.01% detergent, some toxic metals, *e.g.* cadmium at about 5 mg kg^{-1}, and residuals of pesticides from time to time. The sewage sludges produced in urban areas in industrial regions are an even worse mixture, the concentrations of toxic metals, pesticides, and biocides being particularly undesirable in the context of sludge disposal at sea. It is for this reason, rather than any other, that the sea disposal of sewage sludge is to cease within the EC by the turn of the century. The pretreatment of sludge is geared essentially to reducing the bulk of the sludge before disposal to land, sea, or incineration. Anaerobic digestion of sludge to yield methane for power generation reduces the bulk of the solids content of sludges but this does not give great advantage when the sludge contains only 3% solids. However, aeration of sludge after digestion followed by a quiescent period results in substantial separation of water, which if drawn off can give a 20% reduction is sludge volume. There is a variety of machinery available for the dewatering of sludge to yield a handleable sludge cake for use as a fertilizer, for landfill, or for incineration. All are based on chemical conditioning followed by centrifuging, vacuum filtration, or plate or filter belt pressing.

Treatment of Industrial Wastewaters

Industrial wastes can be divided into four categories, namely :

1. Wastewaters that have been changed in chemical quality during use in, or in connection with, manufacturing processes. This category consists of process waters and unclean cooling waters.
2. Contaminated run-off from roofs and yards at industrial premises.
3. Clean cooling waters.
4. Grossly contaminated wastewaters, waste chemicals, and liquid or semiliquid sludges of small volume kept separate, or separated from, the main wastewater streams of manufacturing processes. These wastes are usually transported away for specialist disposal.

Table 6.10 Typical limits and pre-treatments given to industrial wastewaters before discharge to sewers

Contaminant	*Consent limit* ($mg\ L^{-1}$)	*Pre-treatment (where necessary)*
Suspended solids	400—1000	Screening and settlement
BOD	500—1000	High rate bio-oxidation
Oil and grease	10	Passage through oil traps or separators
Cyanide	1—5	Chlorination or enzyme treatment for CN removal
Heavy metals	1—20	Alkaline precipitation and settlement
Acidity/alkalinity	pH 10 and 5	Neutralization
Solvents	Substantially absent	Recovery or activated carbon treatment
Strong dyestuff colour	Low colour	Bleaching with chlorine

Clean cooling waters running to waste should as far as possible be recycled repeatedly and the bleed-off run to disposal with the main process stream. The remaining polluting waters in categories 1 and 2 are best disposed of, if practicable, to public sewers after such pre-treatment as may be necessary. Table 6.10. gives further information on and methods of industrial wastewater treatment.

CHAPTER 7

The Ocean Environment

Introduction

Seawater contains dissolved gases, and as a consequence is both well-oxygenated, although exceptional environments exist, and buffered at a pH of about 8. There are electrolytic salts, the ionic strength of seawater being approximately 0.7, and multitudinous organic compounds in solution. At the same time, there is a wide range of inorganic and organic particles in suspension. These comfortable distinctions become quite confused in seawater. Some molecules present in true solution are sufficiently large to be retained by a filter. Surface adsorption accumulates organic coatings and scavenges dissolved elements. Some elements, particularly those with biochemical functions, may be rapidly removed from solution. Concurrently, reactions involving geological time scales are proceeding slowly. Yet despite this apparent complexity, many aspects of the composition of seawater and chemical oceanography can now be explained with recourse to the fundamental principles of chemistry. This chapter serves to bridge the gap between those with environmental expertise and those with a traditional chemical background.

The Ocean as a Biogeochemical Environment

A traditional approach utilized in geochemistry, and now also in environmental chemistry, is to consider the system under investigation as a reservoir. For a given component, the reservoir has sources (inputs) and sinks (outputs). The system is said to be at equilibrium, or operating under steady-state conditions, when a mass balance between inputs and outputs is achieved. An imbalance could signify that an important source or sink has been ignored. Alternatively, the system may be perturbed, possibly anthropogenically mediated, and therefore changing towards a new equilibrium state.

Processes within the reservoir that affect the temporal and spatial distribution of a given component are transportation and

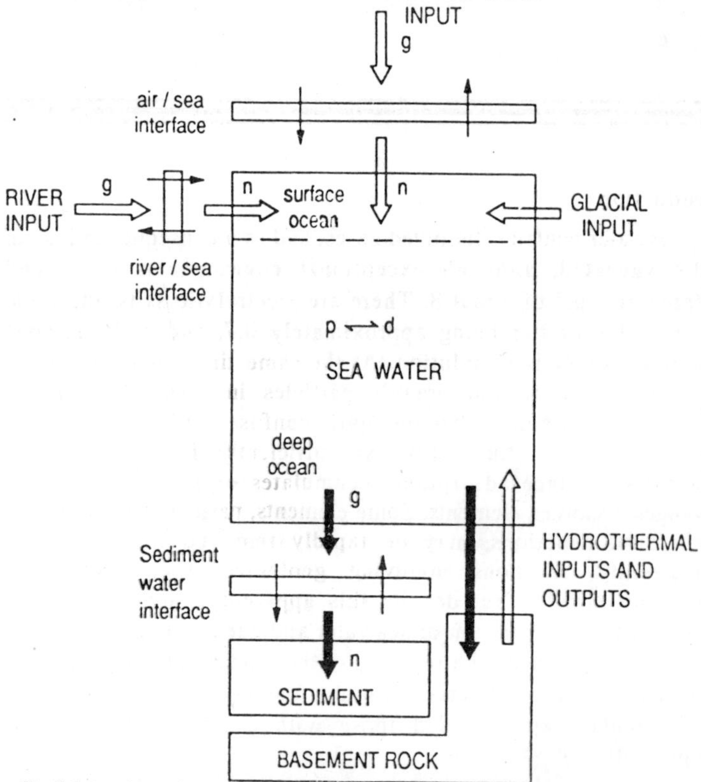

Fig.7.1 *A schematic representation of the ocean reservoir. The source and sink fluxes are designated as g and n, referring to gross and net fluxes, thereby indicating that interactions within the boundary regions can modify the mass transfer. Within seawater, the p ⇌ d term signifies that substances can undergo particulate–dissolved interactions. However, it must be appreciated that several transportation and transformation process might be operative*

Source: R. Chester,'Marine Geochemistry',Chapman and Hall, London, 1990, pp. 698.

transformations. The physics and biology within the system play a role. Clearly transport effects are dominated by the hydrodynamic regime. Although transformations could involve chemical (dissolution, redox reactions, speciation changes) or geological (sedimentation) processes, biological activity can control nutrient and trace metal distributions. Furthermore, the biota influence concentrations of O_2 and CO_2 which in turn determine the pH and *pe* (*i.e.* the redox potential see later), respectively. For these reasons, some fundamental aspects of descriptive physical and biological oceanography are included in this chapter.

In terms of biogeochemical cycling, the ocean constitutes a large reservoir. The surface area is 361.11×10^6 km^2, nearly 71% of the earth's surface. The average depth is 3.7 km, but depths in the submarine trenches can exceed 10 km. The ocean contains about 97% of the water in the global hydrological cycle. A schematic representation of the ocean reservoir is presented in Fig.7.1. The material within it can be operationally defined, usually on the basis of filtration, as dissolved or particulate. The ocean is divided into two layers, with distinct surface and deep waters. The boundary regions are also distinguished as the composition in these regions can be quite different to bulk seawater. Furthermore, interactions within these environments can alter the mass transfers across the boundary.

Material supplied to the ocean originates from the atmosphere, rivers, glaciers, and hydrothermal waters. The relative importance of these pathways depends upon the component considered and geographic location. River run-off generally constitutes the most important source. Transported material may be either dissolved or particular, but discharges are into surface waters and confined to coastal regions. Hydrothermal waters and vents are associated with seafloor spreading ridges. Seawater can circulate into the fissured rock matrix and come into contact with newly forming basalt. Compositional changes in the aqueous phase occur due to seawater-rock interactions and the release of material from the mantle into solution. Thus, hydrothermal activity released dissolved components into the ocean at great depths. This is an important source of some elements, such as Li, Rb, and Mn. The atmosphere supplies particulate material globally to the surface of the ocean. This is the most prominent pathway for Pb to the world ocean. Aeolian transport is greatest in low latitudes and the Sahara Desert is known to act as an important source of dust. Conversely, glacial activity makes little impact on the world ocean. Glacier-derived material tends to be comprised of physically weathered rock residue, and so relatively insoluble, and also the input is largely confined to polar regions, with Antarctica responsible for approximately 90% of the material. Sedimentation acts as the major removal process. However, volatilization and subsequent evasion to the atmosphere can be important for elements such as Se and Hg that undergo bioalkylation.

Some definitions facilitate the interpretation of chemical phenomena in the ocean. Conservative behaviour signifies that the concentration of a constituent (or absolute magnitude of a property) varies only due to mixing processes. Components or parameters that behave in this manner can be used as conservative indices of mixing. Examples are salinity and potential temperature, the definitions for which are presented in subsequent sections. In contrast, non-conservative behaviour indicates that the concentration of a constituent may vary as a result of biological or chemical processes. Examples of parameters that behave non-conservatively are dissolved oxygen and pH. Residence time, τ, is defined as:

$$\tau = \frac{A}{(dA/dT)}$$

Where A is the total amount of constituent A in the reservoir and dA/dT can be either in rate of supply or the rate of removal of A. This represents the average life time of the component in the system and is, in effect, a reciprocal rate constant. Finally, the photic zone refers to the upper surface of the ocean in which photosynthesis can occur. This is typically taken to be the layer down to the depth at which sunlight radiation has declined to 1% of the magnitude at the surface.

Properties of Seawater

Water is unique substance, with unusual attributes as a result of its structure. The molecule consists of a central oxygen atom with two attached hydrogen atoms forming a bond angle of about 105°. As oxygen is more electronegative than hydrogen, it attracts the shared electrons to a greater extent. Also, the oxygen atom has a pair of lone orbitals. The overall effect produces a molecule with a strong dipole moment, that is, having distinct negative (O) and positive (H) ends. While there are several important consequences, two will be considered here. Firstly, the positive H atoms of one molecule are attracted towards the negative O atoms in adjacent molecules giving rise to hydrogen bonding. This has important implications with respect to a number of physical properties, especially those relating to thermal characteristics. Secondly, the large dipole moment ensures that water is a very polar solvent.

Considering firstly the physical properties, water has much higher freezing and melting points than would be expected for a molecule of molecular weight 18. Water has high latent heats of evaporation and fusion. This means that considerable energy is required to stimulate phase changes, the energy being utilized in hydrogen bond rupturing. Moreover, it has a high specific heat and is a good conductor of heat. Consequently, heat transfer in water by advection and conduction gives rise to uniform temperatures. The density of pure water exhibits anomalous behaviour. In ice, O atoms have 4 H atoms oriented about them tetrahedrally. These units are packed together with a hexagonal symmetry. At the freezing point, 0 °C, ice is less dense than water. Heating breaks hydrogen bonds and molecules can achieve slightly closer packing which causes the density to increase. The maximum density occurs at 4 °C, as at higher temperatures thermal expansion compensates for this compression effect. As will be discussed later, seawater differs in this respect. Thus, fresh ice floats on water which in part explains how rivers and lakes can freeze over but remain liquid at depth. With respect to other properties, water has a high surface tension which is manifest in stable droplet formation and has a relatively low molecular viscosity and therefore is quite a mobile fluid.

Water is an excellent solvent. It is extremely polar and generally can dissolve a wider range of solutes and in greater amounts than any other substance. Water has a very high dielectric constant, a measure of the solvents' ability to keep apart oppositely charged ions. The solvation characteristics of individual ions influence their behaviour in solution, *i.e.* in terms of hydration, hydrolysis, and precipitation. Although water exhibits amphoteric behaviour, electrolytic dissociation is quite small. Furthermore, dissociation gives equal ion concentrations of both H_3O^+ and OH^- and so pure water is neutral. The amphoteric behaviour enhances dissolution of introduced particulate matter through surface hydrolysis reactions.

While the concept will be considered in detail below, the term salinity (*S%*) is introduced here as a measure of the salt content of sea water, expressed in units g kg^{-1}. A typical value for oceanic waters is 35 g kg^{-1}. In an oceanographic context, the most important consequence of the addition of salt to water is the effect on density. However, many of the characteristics outlined above are also altered. The addition of

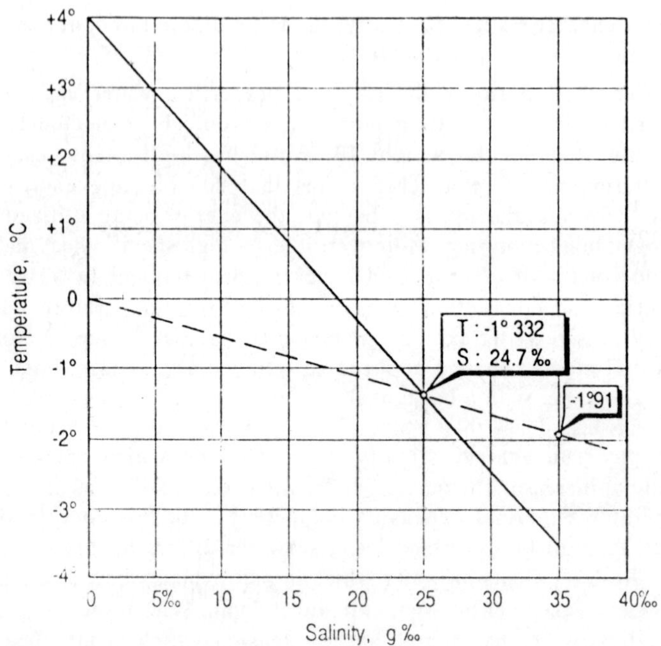

Fig. 7.2 *The temperature of maximum density (—) and freezing point (– –) of seawater as a function of dissolved salt content*

electrolytes can cause a small increase in the surface tension. This effect is not generally observed in seawater due to the presence of surfactants, which decrease the surface tension and so facilitate foam formation. As illustrated in Fig.7.2, the presence of salt does depress the temperature of maximum density and the freezing point of the solution relative to pure water. Thus, seawater with a typical salt content of 35 g kg^{-1} freezes at approximately —1.9 ° C and the resulting ice is more dense than the solution. As a further consequence, the freezing process tends to produce fresh ice overlying a more concentrated brine solution. Salts can be precipitated at much lower temperatures, *i.e.* mirabilite ($Na_2SO_4.2H_2O$) at —8.2 ° C and halite (NaCl) at —23°C. Some brine inclusions and salt crystals can become incorporated into the ice.

The fundamental properties of seawater are temperature and salinity. Together with the pressure (*i.e.* depth dependent), these parameters control the density of the water. The density in turn determines the buoyancy of the water and pressure gradients. Small density differences integrated over oceanic scales cause considerable pressure gradients and result in currents.

Surface water temperatures are extremely variable, obviously influenced by location and season. The minimum temperatures found in polar latitudes are almost —2 °C. Equatorial waters can reach 30 °C. It would be expected that the temperature decreases with depth; easily explained in that water density decreases with temperature and so, in the absence of other stabilizing influences, elevated temperatures at depth would produce a buoyant mass resulting in mixing. Temperature variations with depth are far from consistent. Regions in which mixing is prevalent, *i.e.* especially in the surface waters, produces a layer in which the temperature is relatively constant. The zone immediately beneath would normally exhibit a sharp change in temperature, known as the thermocline. The thermocline in the ocean extends down to about 1000 m within equatorial and temperate latitudes. It acts as an important boundary in the ocean, separating the surface and deep layers, and limiting mixing between these two reservoirs.

Below the thermocline, the temperature changes only little with depth. The temperature in seawater is non-conservative because adiabatic compression causes a slight increase in the *in situ* temperature measured at depth. For instance in the Mindanao Trench in the Pacific Ocean, the temperature at 8500 m and 10000 m is 2.23 °C and 2.48 °C, respectively. The term potential temperature is defined to be the temperature that the water parcel would have if raised adiabatically to the ocean surface. For the examples above, the potential temperatures are 1.22 °C and 1.16 °C, respectively. Potential temperature is a conservative index.

Salinity in the surface waters in the open ocean range between 33 and 37% the main control being the balance between evaporation and precipitation. The highest salinities occur in regional seas where the evaporation rate is extremely high, namely the Mediterranean Sea (38—39‰) and the Red Sea (40—41‰). Within the world ocean, the salinity is greatest in latitudes of about 20° where the evaporation exceeds precipitation. Lower salinities occur poleward as evaporation diminishes and near the equator where precipitation is very high. Local effects can be important, as evident in the vicinity of large riverine discharges which dilute the salinity. Salinity variations with depth are related to the origin of the deep waters and so will be considered in the section on oceanic circulation. A zone in which the salinity exhibits a marked gradient is known as a halocline.

Whereas the density of pure water is 1.000 g ml^{-1}, the density of sea water ($S‰ = 35‰$) is about 1.03 gml^{-1}. The term 'sigma-tee', σ_τ is used to denote the density (actually the specific gravity and hence a dimensionless number) of water at atmospheric pressure based on temperature and salinity *in situ*. Density increases, and so the buoyancy decreases, with an increase in σ_τ. It is defined as:

$$\sigma_\tau = (\text{specific gravity}_{S‰, T}{}^{-1}) \times 1000$$

In a plot of temperature against salinity (a*T* — *S* diagram), constant σ_τ appear as curved lines which denote waters of constant pressure and are known as isopycnals. A zone in which the pressure changes greatly is known as a pycnocline. Within the water column, a pycnocline therefore separates waters with distinctive temperature and salinity characteristics, generally indicative of different origins. A *T*—*S* diagram can also be used to estimate the properties resulting from the mixing of two water masses. As noted above, the temperature is not a conservative property, and therefore σ_τ is also non-conservative. To circumvent the associated difficulties of interpretation, an analogous term known as the potential density, σ_0, is defined on the basis of potential temperature (*i.e.* temperature if adjusted to a pressure of 1 atm.) instead of *in situ* temperature. The σ_0 is therefore a conservative index.

Salinity Concepts

Salinity is a measure of the salt content of seawater. Developments in analytical chemistry have led to an historical evolution of the salinity concept. Intrinsically it would seem to be a relatively straight forward task to measure. This is true for imprecise determinations which can be quickly performed using hand-held refractometers. The salinity affects seawater density, and thus, the impetus for high precision in salinity measurements came from physical oceanographers.

The first techniques utilized for the determination of salinity, involving the gravimetric analysis of salt left after evaporating seawater to dryness, were fraught with difficulties. Variable amounts of water of crystallization might be retained. Some salts, such as $MgCl_2$ can decompose leaving residues of uncertain composition.

Other constituents, especially organic material, might be volatilized or oxidized. Overall, such methods led to considerable inconsistencies and inaccuracies.

Chlorinity ($Cl\%_0$) is the chloride concentration in seawater, expressed as g kg^{-1}, as measured by Ag^+ titration (*i.e.* assuming Cl^- to be the only reactant). The relationship of interest was that between $S\%_0$ and $Cl\%_0$, given as:

$$S\%_0 = 1.805Cl\%_0 + 0.030$$

As a calibrant solution for the $AgNO_3$ titrant, Standard Seawater was prepared that had certified values for both Chlorinity and salinity. This salinity—Chlorinity relationship was derived on the basis of only 9 seawater samples that were somewhat atypical, and it has been redefined in recent years using a much larger set of samples representative of oceanic waters to become:

$$S\%_0 = 1.80655\ Cl\%_0$$

The third category of salinity methodologies was based on conductometry. These continue to be the most widely used methods, because electrical conductivity measurements can provide salinity values with a precision of $\pm 0.001\%_0$. The conductivity of a solution is proportional to the salt content. High precision requires temperature control of samples and standards to within ± 0.001 °C. Standard Seawater, now also certified with respect to conductivity, provides the appropriate calibrant solution. Application of a non-specific technique like conductometry relies upon the assumption that the sea salt matrix is invariant, both spatially and temporally. Thus, the technique cannot be reliably employed in marine boundary environments where the sea water composition differs to the bulk characteristics. There are two types of procedures commonly used. Firstly, a Wheatstone Bridge circuit can be set up whereby the ratio of the resistance of unknown seawater to standard seawater balances the ratio of a fixed resistor to a variable resistor. The system uses alternating current to minimize electrode fouling. Alternatively, the conductivity can be measured by magnetic induction, in which case the sensor consists of a plastic tube containing sample seawater that links two transformers. An oscillator establishes a current in one transformer which induces current flow within the tube, the magnitude of which depends upon the salinity of the sample. This in turn induces a current in the second transformer which can then be measured. This design has been exploited for *in situ* conductivity measurements.

Oceanic Circulation

The distribution of components within the ocean is determined by both transportation and transformation processes. A brief outline of oceanic circulation is necessary to ascertain the relative influences. Two main flow systems must be considered. Surface circulation is established by the prevailing wind patterns but modified by Coriolis force and deep circulation is determined by gravitational forces. The Coriolis force is the acceleration due to the earth's rotation. It acts to deflect moving fluids (*i.e.* both air and water) to the right in the northern hemisphere and to the left in the southern hemisphere. The magnitude of the effect is a function of latitude, being nil at the equator and increasing polewards.

The surface circulation is restricted to the upper layer influenced by the wind, typically about 100 m. However, underlying water can be transported up into this zone when horizontal advection is insufficient to maintain the superimposed flow fields. This process is called upwelling and is of considerable importance in that biochemical respiration of organic material at depth ensures that the ascending water is nutrient-rich. Upwelling occurs in the eastern oceanic boundaries where longshore winds result in the offshore transport of the surface water. Examples are found off Peru and West Africa. Similar processes cause upwelling off Arabia, but this is seasonal due to the monsoon effect. A divergence is a zone in which the flow fields separate. In such a case, upwelling may result as observed in the equatorial Pacific. It should be noted that a region in which the stream lines come together is known as a convergence, and water sinks in this zone.

The deep circulation is controlled by the density of the water. If the density of a water body increases, it has a tendency to sink. Subsequently it will spread out over a horizon of uniform σ_0. As the density can be raised due to either an increase in the salinity or a decrease in the temperature, the deep water circulatory system is also known as thermohaline circulation. The most dense waters are formed in polar regions due to the relatively low temperatures and the salinity enhancement that results from sea ice formation. Antarctic Bottom Water (ABW) is generated in the Weddell Sea and flows northward into the South Atlantic. North Atlantic Deep Water (NADW) is formed in the Norwegian Sea and off the southern coast of Greenland. The flow of the NADW can be traced southwards through the Atlantic Ocean to Antarctica. It is diverted eastward into the Southern Indian Ocean and South Pacific. There it heads northwards and either enters the North Pacific or becomes mixed upward into the surface layer in the equatorial region. The transit time is on the order of 1000 years. As noted previously, the thermocline acts as an effective barrier against mixing of dissolved components in the ocean. Consequently, this deep water formation process in high latitudes is important because it facilitates the relatively rapid transport of material from the surface of the ocean down to great depths. The deep advection of atmospherically derived CO_2 is a pertinent example.

Intermediate waters within the water column can be formed by diverse processes. In the southern South Atlantic, the NADW overrides the more dense ABW. Antarctic Intermediate Water results from water sinking along the Antarctic Convergence (~ 50° S). Relatively warm, saline water exits the Mediterranean Sea at depth and can be identified as a distinctive layer within the North and South Atlantic.

SEAWATER COMPOSITION AND CHEMISTRY

Major Constituents

The major constituents in seawater are conventionally taken to be those elements present in typical oceanic water ($35\ g\ kg^{-1}$) at concentrations greater than $1 mg\ kg^{-1}$, excluding Si which is an important nutrient in the marine environment. The concentrations and main species of these elements are presented in Table 7.1.

Table 7.1 Chemical species and concentrations of the major elements in seawater

Element	*Chemical species*	Concentration for S = 35‰ ($mol\ dm^{-3}$)	($g\ kg^{-1}$)
Na	Na^+	4.79×10^{-1}	10.77
Mg	Mg^{2+}	5.44×10^{-2}	1.29
Ca	Ca^{2+}	1.05×10^{-2}	0.4123
K	K^+	1.05×10^{-2}	0.3991
Sr	Sr^{2+}	9.51×10^{-5}	0.00814
Cl	Cl^-	5.59×10^{-1}	19.353
S	SO_4^{2-}, $NaSO_4^-$	2.89×10^{-2}	0.905
C (inorganic)	HCO_3^-, CO_3^{2-}	2.35×10^{-3}	0.276
Br	Br^-	8.62×10^{-4}	0.673
B	$B(OH)_3$, $B(OH)_4^-$	4.21×10^{-4}	0.0445
F	F^-, MgF^+	7.51×10^{-5}	0.00139

Source : *D. Dyrssen and M. Wedborg, 'The Sea', ed. E. Goldberg, John Wiley and Sons, New York, 1974, p.181.*

The residence times for some elements are presented in Table 7.2. The major constituents generally have long residence times. The residence time is a crude measure of a constituent's reactivity in the reservoir. The aqueous behaviour and rank ordering can be appreciated simply in terms of the ionic potential given by the ratio

Table 7.2 The residence time and speciation of some elements in the ocean

Element	*Principal species*	*Concentration* ($mol\ l^{-1}$)	*Residence time* *(years)*
Li	Li^+	2.6×10^{-5}	2.3×10^6
B	$B(OH)_3$, $B(OH)_4^-$	4.1×10^{-4}	1.4×10^7
F	F^-, MgF^+	6.8×10^{-5}	5.2×10^5
Na	Na^+	4.68×10^{-1}	6.8×10^7
Mg	Mg^{2+}	5.32×10^{-2}	1.2×10^7
Al	$Al(OH)_4^-$, $Al(OH)_3$	7.4×10^{-8}	1.0×10^2
Si	$Si(OH)_4$	7.1×10^{-5}	1.8×10^4
P	HPO_4^{2-}, PO_4^{3-}, $MgHPO_4$	2×10^{-6}	1.8×10^5
Cl	Cl^-	5.46×10^{-1}	1×10^8
K	K^+	1.02×10^{-2}	7×10^6
Ca	Ca^{2+}	1.02×10^{-2}	1×10^6
Sc	$Sc(OH)_3$	1.3×10^{-11}	4×10^4
Ti	$Ti(OH)_4$	2×10^{-8}	1.3×10^4
V	$H_2VO_4^-$, HVO_4^{2-}, $NaVO_4^-$	5×10^{-8}	8×10^4
Cr	CrO_4^{2-}, $NaCrO_4^-$	5.7×10^{-9}	6×10^3
Mn	Mn^{2+}, $MnCl^+$	3.6×10^{-9}	1×10^4
Fe	$Fe(OH)_3$	3.5×10^{-8}	2×10^2
Co	Co^{2+}, $CoCO_3$, $CoCl^+$	8×10^{-10}	3×10^4
Ni	Ni^{2+}, $NiCO_3$, $NiCl^+$	2.8×10^{-8}	9×10^4
Cu	$CuCO_3$, $CuOH^+$, Cu^{2+}	8×10^{-9}	2×10^4
Zn	$ZnOH^+$, Zn^{2+}, $ZnCO_3$	7.6×10^{-8}	2×10^4
Br	Br^-	8.4×10^{-4}	1×10^8
Sr	Sr^{2+}	9.1×10^{-5}	4×10^6
Ba	Ba^{2+}	1.5×10^{-7}	4×10^4
La	La^{3+}, $LaCO_3^+$, $LaCl^{2+}$	2×10^{-11}	6×10^2
Hg	$HgCl_4^{2-}$	1.5×10^{-10}	8×10^4
Pb	$PbCO_3$, $Pb(CO_3)_2^{2-}$, $PbCl^+$	2×10^{-10}	4×10^2
Th	$Th(OH)_4$	4×10^{-11}	2×10^2
U	$UO_2(CO_3)_2^{4-}$	1.4×10^{-8}	3×10^6

Source: P. Brewer, 'Chemical Oceanography' ed. J.P. Riley and G. Skirrow, Academic Press, London, 2nd Ed., 1975, Vol. I., p.415.
K.W. Bruland, 'Chemical Oceanography' ed. J.P. Riley and R. Chester, Academic Press London, 1983, Vol.8., p. 157.

of electronic charge to ionic radius (Z / r). Elements with $Z/r < 3$ are strongly cationic. The positive charge density is relatively diffuse, but sufficient to attract and orientate an envelope of water molecules forming a hydrated cation. As the ionic potential increases, the force of attraction towards the water similarly rises to the extent that one oxygen—hydrogen bond in the molecule breaks. This causes the solution pH to fall the metal hydroxides to form. Neutral hydroxides tend to be relatively insoluble and so precipitate. However, in the more extreme case for which $Z/r > 12$, the attraction toward the oxygen is so great that both bonds in the associated water molecules are broken. The reaction product is an oxyanion, usually quite soluble because of the associated water molecules are broken. The reaction product is an oxyanion, usually quite soluble because of the associated anionic charge. Thus in seawater, those elements (Al, Fe) having a tendency to form insoluble hydroxides have short residence times. This is also true for elements that exist preferentially as neutral oxides (Mn, Ti). Hydrated cations (Na^+, Ca^{2+}) and strongly anionic species (Cl^-, Br^-, $UO_2(CO_3)_2^{4-}$) have long residence times. This treatment is, of course, somewhat of an over-simplification ignoring the rather significant role that biological organisms play in nutrient and trace element chemistry.

Dissolved Gases

Gas Solubility and Air-Sea Exchange Processes

The ocean contains a vast array of dissolved gases. Some of the gases such as Ar and chlorofluorocarbons behave conservatively and can be utilized as tracers for water mass movements and ventiiation rates. Equilibrium processes at the air-sea interface generally lead to saturation, and then this concentration remains unchanged once the water sinks. Thus, the gas concentration is characteristic of the lost contact with the atmosphere. Deep waters usually contain no CFCs as such anthropogenic compounds have only a recent history of use. There are several important non-conservative gases, which exhibit wide variations in concentration due to biological activity. O_2 determines the redox potential in seawater and CO_2 buffers the ocean at pH 8. Ocean—atmosphere exchange processes for gases such as CO_2 and dimethyl sulphide may play an important role in climate change.

The solubility of gases in water is affected by both temperature and salinity. The trends are such that gas solubility increase with a decrease in temperature or an increase in salinity. The changes in solubility are non-linear and differ dramatically for various gases.

At the ocean—atmosphere interface, exchange of gases occurs to achieve an equilibrium between the two systems and as a result, gases become saturated. However, super-saturation can be achieved by several mechanisms. Firstly, bubbles that form from white cap activity can be entrained and dissolved at depth. Secondly if two water masses that have been equilibrated at different temperatures are mixed, then the resulting water body would be super-saturated. Thirdly, gases that are produced *in situ* by biological activity may become super-saturated, particularly when evasion to the atmosphere is not favoured.

Table 7.3 The net global fluxes of some trace gases across the air/sea interface (from Chester, 1990)1

Gas	*Global air-sea direction*[a]	*Flux magnitude*[b]
CH_4	+	$10^{12} - 10^{13}$
Man—made CO_2	–	6×10^{15}
N_2O	+	6×10^{12}
CCl_4	–	10^{10}
CCl_4	=	~0
CCl_3F	–	5×10^9
CCl_3F	=	~0
CH_3I	+	$3 - 13 \times 10^{11}$
CO	+	$100 \pm 90 \times 10^{12}$
I_2	+	$4 \pm 2 \times 10^{12}$
Hg	+	$\sim 2 \times 10^9$

[a] + indicates sea → air flux direction, — indicates air → sea flux direction, = indicates no net flux [b] Units in g (of the compound, where applicable) year $^{-1}$

Air–sea exchange processes are dependent upon the concentration gradient and the transfer velocity. The transfer velocity is not a constant, but rather depends upon several physical parameters such as temperature, wind speed, and wave state. The exchange can also be attenuated by the presence of a surface film or slick. Alternatively, the exchange can be facilitated by bubble formation. The concentration gradient determines the direction of the flux, into or out of the ocean. Net global fluxes fc- some gases are presented in Table 7.3. The atmosphere serves as the source of material for conservative gases, especially those of anthropogenic origin, but several gases produced *in situ* by biological activity evade from the ocean.

Oxygen

Oxygen is a non-conservative gas. The concentration varies throughout the water column, its distribution being greatly influenced by biological activity. The generalized chemical equation for carbon fixation is often given as:

$$_{n}CO_2 + {}_{n}H_2O \Longleftrightarrow (CH_2O)_n + {}_{n}O_2$$

During photosynthesis this reaction proceeds to the right, thereby producing organic material, as designated by $(CH_2O)_n$, and O_2. The surface waters become equilibrated with respect to atmospheric O_2. However, it is possible for these waters to become super-saturated with O_2 during periods of intense photosynthetic activity. Respiration occurs as the above reaction proceeds to the left and O_2 is consumed. Photosynthesis is obviously restricted to the upper ocean (in the photic zone) and generally exceeds respiration. However, the relative importance of the two processes changes with depth. The oxygen compensation depth is the horizon in the water column at which the rate of O_2 production by photosynthesis equals the rate of respiratory O_2 oxidation.

Below the photic zone, O_2 is utilized in chemical and biochemical oxidation reactions. The concentration diminishes with depth to develop an oxygen minimum zone. Thereafter, the O_2 concentration in deeper waters begins to increase because these waters originated from polar regions. They were cold and in equilibrium with atmospheric gases at the time of sinking, but subsequently lost little of the dissolved O_2 because the flux of organic material to deep waters is relatively small.

The dissolved O_2 content of seawater has a significant control on the redox potential, often designated in environmental chemistry by *pe*. This is defined with reference to electron activity in an analogous fashion to pH and thus:

$$pe = -\log\{e^-\}$$

The relationship between *pe* and the more familiar electrode potential *E* is:

$$pe = \frac{F}{2.303\,RT}\,E$$

and for the standard state:

$$pe^{\ominus} = \frac{F}{2.303RT}E^{\ominus}$$

Whereas a high value of Pe indicates oxidizing conditions, a low value signifies reducing conditions. Oxygen plays a role via the reaction:

$$O_2 + 4H^+ + 4e^- \rightleftharpoons 2H_2O$$

At 20°C,$K = 10^{8.1}$, and so water of pH = 8.1 in equilibrium with atmospheric O_2 ($_pO_2$ = 0.21 atm) has *pe* = 12.5. This conforms to surface conditions but the *pe* decreases as the O_2 content diminishes with depth. The oxygen minimum is particularly well developed beneath the highly productive surface waters of the eastern tropical Pacific Ocean. There is a large flux of organic material to depth and subsequent oxidation. The O_2 becomes sufficiently depleted that the resulting low redox conditions causes NO_3^- to be reduced to NO_2^-. In some special environments characterized by poor flushing and hence stagnant water, the oxygen may be completely utilized producing anoxic conditions. Such regions represent atypical marine environments where reducing conditions prevail.

O_2 can be used as a tracer to help identify the origin of water masses. The warm, saline intrusion in to the Atlantic Ocean from the Mediterranean Sea is relatively O_2 deficient. Alternatively, the waters descending from polar regions have elevated O_2 concentrations.

Carbon Dioxide and Alkalinity

Marine chemists sometimes adopt activity conventions quite different to those traditionally used in chemistry. It is useful to preface a discussion on the carbon dioxide—Calcium carbonate system in the oceans with a brief outline of pH scales. Although originally introduced in terms of ion concentration, today the definition of pH is based on hydrogen ion activity and is:

$$pH = -\log a_H$$

where a_H refers to the relative hydrogen ion activity (*i.e.* dimensionless as is pH).

The biogeochemical cycle of inorganic carbon in the ocean is extremely complicated. It involves the transfer of gaseous carbon dioxide from the atmosphere into solution. Not only is this a reactive gas that readily undergoes hydration in the ocean, but also it is fixed as organic material by marine phytoplankton. Several marine organisms utilize calcium carbonate to form shells. Surface waters are super-saturated with respect to aragonite and calcite, forms of $CaCo_3$ but precipitation is limited to coastal lagoons such as found in the Bahamas. Inorganic carbon removed from surface waters by biological processes can be regenerated at depth. This may result through either the respiratory oxidation of organic material or the dissolution of shells in the undersaturated waters found at depth. Nonetheless, calcitic oozes of biogenic origin constitute a major component in marine sediments. Finally, the inorganic carbon equilibrium is responsible for buffering seawater at a pH near 8 on time scales of centuries to millennia.

There are several equilibria to be considered. Firstly, CO_2 is exchanged across the air—sea interface:

$$CO_{2(g)} \rightleftharpoons CO_{2(aq)}$$

The equilibrium process obeys Henry's Law, but dissolved CO_2 reacts rapidly with water to become hydrated as:

$$CO_2 + H_2O \rightleftharpoons H_2CO_3$$

Relative to the exchange process, the hydration reaction forming carbonic acid occurs quite quickly. This means that the concentration of dissolved CO_2 is extremely low. The two processes can be considered together as:

$$CO_2 + H_2O \rightleftharpoons H_2CO_3$$

The equilibrium constant is then:

$$K_{CO_2} = \frac{\{H_2CO_3\}}{p_{CO_2}\{H_2O\}}$$

where p_{CO_2} is the partial pressure of CO_2 in the marine troposphere. Carbonic acid undergoes dissociation:

$$H_2CO_3 + H_2O \rightleftharpoons H_3O^+ + HCO_3^-$$

$$HCO_3^- + H_2O \rightleftharpoons H_3O^+ + CO_3^{2-}$$

for which the first and second dissociation constant (using $\{H^+\}$ rather than $\{H_3O^+\}$) are:

$$K_1 = \frac{\{H^+\}\{HCO_3^-\}}{\{H_2CO_3^-\}}$$

$$K_2 = \frac{\{H^+\}\{CO_3^{2-}\}}{\{HCO_3^-\}}$$

The hydrogen ion activity can be established with a pH meter. However as discussed above, this measurement must be operationally defined. On the other hand, the individual ion activities of bicarbonate and carbonate ions cannot be measured. Instead, ion concentrations are determined, as outlined below, by 'itration. Accordingly, the equilibrium constants are redefined in terms of concentrations. These are then known as apparent rather than true equilibrium constants and distinguished using a prime notation. It must be appreciated that apparent equilibrium constant are not invariant, but rather are affectedly by temperature, pressure, salinity, and, as outlined previously, the pH scale adopted. The apparent dissociation constant are:

$$K'_1 = \frac{\{H^+\}[HCO_3^-]}{[H_2CO_3^-]}$$

$$K'_2 = \frac{\{H^+\}\{CO_3^{2-}\}}{[HCO_3-]}$$

It should be noted that whereas ion activities are denoted by curly brackets { }, concentrations are designated by square brackets [ ]. Analogous to the pH, pK conventionally refers to —long K. The speciation for carbonic acid at 15 °C as a function of pH in seawater equilibrated with atmospheric carbon dioxide (~ 3.5×10^{-4} atm) is given in Fig. 7.3. While there are several confounding features, the pH of seawater can be considered to be buffered by the bicarbonate:

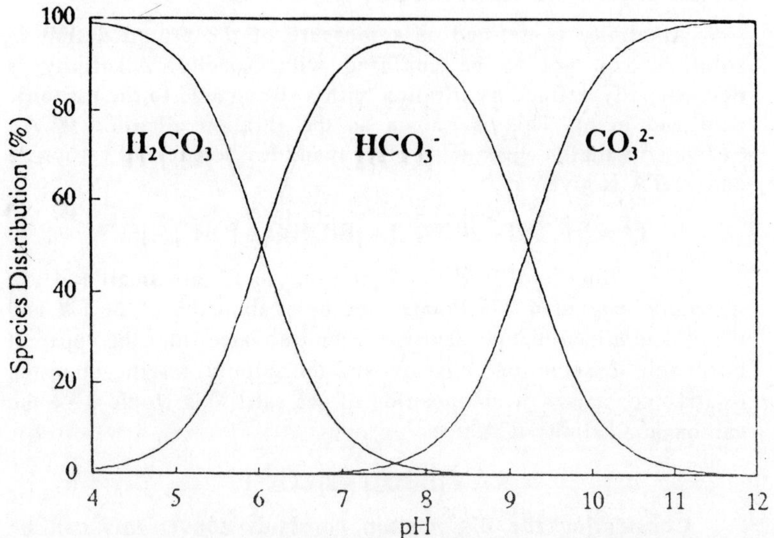

Fig. 7.3 *The distribution of carbonic acid species in seawater of 35% at 15 °C as a function of pH*

carbonate pair. The pH is generally about 8, but is sensitive to the concentration ratio $[HCO_3^-] : [CO_3^{2-}]$ as evident from rearranging the expression for K'_2 to become:

$$\{H^+\} = K'_2 \frac{[HCO_3^-]}{[CO_3^{2-}]}$$

To understand the response of the oceanic CO_2 system to *in situ* biological activity or enhanced CO_2 concentrations in the atmosphere, it is necessary to consider in more detail the factors influencing the inorganic carbon cycle. Two useful parameters can be introduced. Firstly, the total concentration of inorganic carbon, ΣCO_2, in seawater is:

$$\Sigma CO_2 = [CO_2] + [H_2CO_3] + [HCO_3^-] + [CO_3^{2-}]$$

The first term is negligible and as evident in Fig.7.3, the major species at pH 8 are HCO_3^- and CO_3^{2-}.

Alkalinity is defined as a measure of the proton deficit in solution, and not to be confused with basicity. Alkalinity is operationally defined by titration with a strong acid to the carbonic acid end point. This is known as the titration alkalinity (TA). Seawater contains weak acids other than bicarbonate and carbonate and so TA is given as:

$$TA = [HCO_3^-] + 2[CO_3^{2-}] + [B(OH)_4^-] + [OH^-] - [H^+]$$

The influence of $[OH^-]$ and $[H^+]$ on the TA are small and can generally be ignored. The borate contributes about 3% of the TA, and if not determined independently, can be estimated from the apparent boric acid dissociation constants and the salinity, relying upon the relative constancy of composition of sea salt. This would give the carbonate alkalinity (CA):

$$CA = [HCO_3^-] + 2[CO_3^{2-}]$$

Considering the dissociation constants above, this can be alternatively expressed as:

$$CA = \frac{K'_1[H_2CO_3]}{\{H^+\}} + \frac{2K'_1K'_2[H_2CO_3]}{\{H^+\}^2}$$

or

$$CA = \frac{K'_1K_{CO_2}p_{CO_2}}{\{H^+\}} + \frac{2K'_1K'_2K_{CO_2}p_{CO_2}}{\{H^+\}^2}$$

This equation can be rearranged to give the following quadratic expression that can be solved for the pH:

$$CA\{H^+\}^2 - K'_1K_{CO_2}p_{CO_2}\{H^+\} - 2K'_1K'_2K_{CO_2}p_{CO_2} = 0$$

Using constants at 15 °C and assuming a typical seawater alkalinity of 2.30 meq L^{-1}, the pH of seawater in equilibrium with atmospheric $CO_2 = 3.5 \times 10^{-4}$ atm is calculated to be 8.20.

Consider now the effect of altering the pCO_2 in the water. The alkalinity should not change in response to variations in CO_2 alone because the hydration and dissociation reactions give rise to equivalent amounts of H^+ and anions. CO_2 can be lost by evasion to the atmosphere (a process usually confined to equatorial regions) or by photosynthesis. This causes the ΣCO_2 to diminish and the pH to rise, an effect that can be Quite dramatic in tidal rock pools in which pH may then rise to 9. Conversely, an increase in pCO_2, either by invasion from the atmosphere or release following respiration, prompts an increase in ΣCO_2 and a fall in pH. Thus, the depth profiles of pH would mimic that of O_2 but the ΣCO_2 would exhibit a maximum at the oxygen minimum.

There are further confounding influences, in particular concerning $CaCo_3$.$CaCO_3$ in the form of aragonite or calcite is used by many organisms to form calcareous shells (tests). The shells sink and dissolve when the organism dies. The solubility is governed by:

$$Ca^{2+} + CO_3^{2-} \rightleftharpoons CaCO_{3_s}$$

Surface waters are super-saturated with respect to $CaCO_3$, but precipitation rarely occurs, possibly due to an inhibitory by Mg^{2+} forming ion pairs with CO_3^{2-}. The solubility of $CaCo_3$ increases with depth, due to both a pressure effect and the decrease in pH following respiratory release of CO_2, with the result that the shells dissolve. This behaviour not only increases the alkalinity but accounts for the non-conservative nature of Ca^{2+} and inorganic carbon in deep waters. The depth at which appreciable dissolution begins is known as the lysocline. At a greater depth, designated as the Carbonate compensation depth (CCD), no calcareous material is preserved. The depths of the lysocline and CCD are influenced by the flux of organic material and shells, and tend to be deeper under high productivity zones.

The CO_2 and $CaCO_3$ systems are coupled in that the pH buffering in the ocean is due to the reaction:

$$CO_2H_2O + CaCO_3 \rightleftharpoons Ca^{2+} + 2HCO_3^-$$

In addition to the effect noted previously, an input of CO_2 promotes the dissolution of $CaCO_3$. The reaction does not proceed to the right without constraint, but rather meets a resistance given by the Revelle factor, R:

$$R = \frac{dp_{CO_2}/p_{CO_2}}{d\Sigma CO_2/\Sigma CO_2}$$

This value is approximately 10, indicating that the ocean is relatively well buffered against changes in ΣCO_2 in response to variations in atmospheric pCO_2. Although the ocean does respond to an increase in the atmospheric burden of CO_2, the time scales involved are quite considerable. The surface layer can become equilibrated on the order of decades, but as the thermocline inhibits exchange into deep waters, the equilibration of the ocean as a whole with the atmosphere proceeds on the order of centuries. The ventilation of deep water by descending water masses in polar latitudes only partly accelerates the overall process.

Nutrients

Although several elements are necessary to sustain life, traditionally in oceanography, 'nutrients' has referred to nitrate, phosphate, and silicate. The rationale for this classification was that analytical techniques had long been available that allowed the precise determination of these constituents despite their relatively low concentrations. They were observed to behave in a consistent manner, but quite differently to the major constituents in seawater.

The distributions of these three nutrients are determined by biological activity. Nitrate and phosphate become incorporated into the soft parts of organisms.

Silicate is utilized by some organisms, particularly diatoms (phytoplankton) and radiolaria (zooplankton), to form siliceous skeletons. Such skeletons consist of an amorphous, hydrated silicate, $SiO_2.nH_2O$, often called opaline silica.

Nutrients behave much like CO_2 and are removed in the surface layer, especially in the photic zone. Thus, concentrations can become quite low, and indeed sufficiently low to limit further photosynthetic carbon fixation. Following the death of the organisms, they sink. The highest concentrations are found where respiration and bacterial decomposition of the falling organic material are greatest, that is at the oxygen minimum. As a consequence of these processes, the nutrients, including silica, are regenerated. Thus, the concentrations in deep waters are much greater than those observed in the surface waters, and account for the fertilizing effect of upwelling. It should be noted that the siliceous remains behave differently than the calcareous shells discussed previously.The oceans are everywhere under-saturated with respect to silica. Silica is preserved to any great extent only in deep-sea sediments associated with the highly productive upwelling zones in the ocean.

Trace Elements

Trace elements in seawater are taken to be those that are present in quantities less than 1 mg L^{-1}, excluding the nutrient constituents. Analytical difficulties are readily comprehensible when it is appreciated that the concentration for some of these elements can be extremely low, *i.e.* a few pg L^{-1} for platinum group metals. Some trace elements, such as Cs^{+}, behave conservatively and therefore absolute concentrations depend upon salinity. More often, elements are non-conservative and their distributions in both surface waters and the water column vary greatly, reflecting the differing source strengths and removal processes in operation. Generalizations regarding residence times cannot be made as biologically active elements are removed from seawater relatively rapidly but conservative constituents and platinum group metals have rather long residence times on the order of 10^5 years.

Considering firstly the distribution in surface waters, several elements exhibit high concentrations in coastal waters in comparison to levels in the centres of oceanic gyres.Typically this arises because the elements originate predominantly from riverine inputs or diffusion from coastal sediments. However, as they are effectively removed from the surface waters in the coastal regions, little material is advected horizontally. Examples of elements that behave in this way are Cd, Cu, and Ni. In contrast, the concentration of Pb, including ^{210}Pb, is greater in the gyres. This results from strong widespread aeolian signal coupled with less effective removal from surface waters in the gyres.

Clearly the removal mechanisms have a significant effect on dissolved elemental abundances. The two major processes in operation are uptake by biota and scavenging by suspended particulate material. In the first instance, the constituent mimics the behaviour of nutrients.

No consistent pattern for depth profiles exist. Conservative elements trend with salinity variations, provided they have no significant submarine sources. Non-conservative elements may exhibit peak concentrations at different depths in oxygenated waters as:

(1) surface enrichment;

(2) maxima at O_2 minimum;

(3) mid-depth maximum not associated with O_2 minimum;

(4) bottom enrichment.

The criteria for an element such as Pb to exhibit a maximum concentration in surface waters are that the only significant input must be at the surface (aeolian supply) and it must be effectively removed from the water column. Constituents such as As, Ba, Cd, Ni, and Zn exhibit nutrient type behaviour. Those elements (*e.g.* Cd) associated with the soft parts of the organism are strongly correlated with phosphate and are regenerated at the O_2 minimum. Elements (*e.g.* Zn) associated with the skeletal material may exhibit a smooth increasing trend with depth. The third case pertains to elements, notably Mn, that have a substantial input from hydrothermal waters. These are released into the ocean from spreading ridges. Ocean topography is such once these waters are advected away from such regions to the abyssal plains, they are then found at

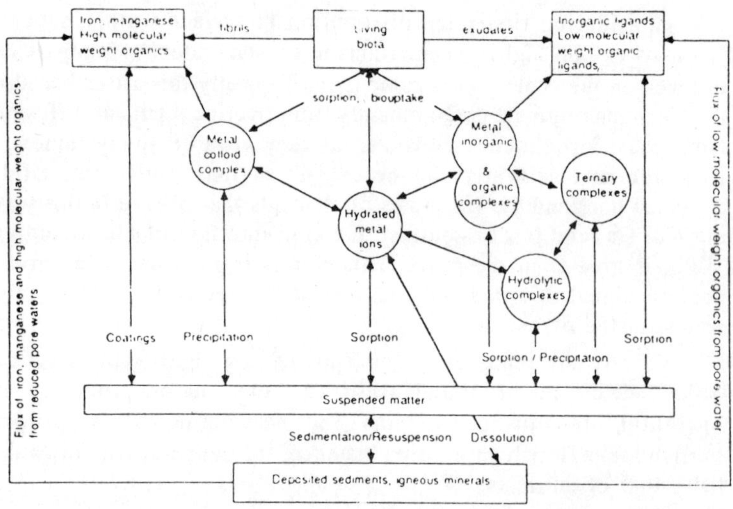

Fig.7.4 *Speciation of metal ions in seawater and the main controlling mechanisms*
Adapted from : L. Ohman and S. Sjoberg,'Metal Speciation: Theory, Analysis and Applications', ed. J.R. Kramer and H.E. Allen, Lewis Publishers, Chelsea, 1988, p.l.

some intermediate depth. Bottom enrichment is observed for elements (*e.g.* Mn) that are remobilized from marine sediments. The behaviour of Al combines features outlined above, resulting in a mid-depth minimum concentration. Surface enrichment evident in mid (41 °N) but not high (~ 60 °*N*) latitudes in the North Atlantic results from the solubilization of aeolian material. Removal occurs via scavenging and incorporation into siliceous skeletal material. Subsequent regeneration by shell dissolution increases deep water Al levels.

Physicochemical Speciation

Physicochemical speciation refers to the various physical and chemical forms in which an element may exist in the system. In oceanic waters, it is difficult to determine speciation directly. Whereas some specific species can be analysed, others can only be inferred from thermodynamic equilibrium models. For instance, this was done with respect to the speciation of carbonic acid in Fig. 7.3. Often an element is fractionated into various forms that behave similarly under a given physical (*e.g.* filtration) or chemical (*e.g.* ion exchange) operation. The resulting distribution of the element is highly dependent upon the procedure utilized, and so known as *operationally defined.*

Physicochemical speciation determines the environmental mobility of an element, especially with respect to partitioning between the water and sediment reservoirs. The influence can be affected through various mechanisms as summarized in Fig 7.4. Settling velocities, and by implication the residence time, are controlled by the size of the particle. Thus, dissolved to particulate interactions involving adsorption, precipitation, or biological uptake can effectively remove a constituent from the water column. The redox state can have a comparable influence. Mn and Fe are reductively remobilized from sediments. Following their release from either hydrothermal sources or interstitial waters, they rapidly undergo oxidation to form colloids which then are quickly removed from the water column. Speciation also determines the bioavailability of an element for marine organisms. It is generally accepted that the uptake of trace elements is limited to free ions and some types of lipid-soluble organic complexes. This is important in that high concentrations of some elements are toxic but, of course, some substances are essential for life.

An element can exist in natural waters in a range of forms that exhibit a size distribution as indicated in Table 7.4. Entities in true solution include ions, ion pairs, complexes, and a wide range of organic molecules which can span several size categories. At the smallest extreme are ions, which exist in solution with a coordinated sphere of water molecules as discussed previously. Na^+, K^+, and Cl^- exist predominantly as free hydrated ions. Collisions of oppositely charged ions occur due to electrostatic attraction, and can produce an ion pair. An ion pair is the transient coupling of a cation and anion, in which each retain its co-ordinated water envelope. While impossible to measure directly, concentrations can be calculated with knowledge of the ion activities and stability constant. For the formation of the ion pair $NaSO_4^-$ via:

$$Na^+ + SO_4^{2-} \rightleftharpoons NaSO_4^-$$

the stability constant is defined as:

$$K = \frac{\{NaSO_4^-\}}{\{Na^+\}\,\{SO_4^{2-}\}}$$

Ion pair formation is important for Ca^{2+}, Mg^{2+}, SO_4^{2-} .and HCO_3^-

Table 7.4 The size distribution of trace metal species in natural waters

Size range	Metal species	Examples	Phase state
< 1nm	Free metal ions	Mn^{2+}, Cd^{2+}	Soluble
1—10 nm	Inorganic ion pairs, inorganic complexes, low molecular weight organic complexes	$NiCl^+$, $HgCl_4^{2-}$ Zn-fulvates	Soluble
10—100 nm	High molecular weight organic complexes	Pb-humates	Colloidal
100—1000 nm	Metal species adsorbed onto inorganic colloids, metals associated with detritus	Co–MnO_2, Pb–$Fe(OH)_3$	Particulate
> 1000 nm	Metals adsorbed into living cells metals adsorbed onto or incorporated into mineral solids and precipitates	Cu–clays, $PbCO_{3s}$	Particulate

Source: S.J. de Mora and R.M. Harrison, 'Hazard Assessment of Chemicals, Current Developments',ed. J. Saxena, Academic Press, London, Vol.3, 1984, p.1.

If the attraction is sufficiently great, a dehydration reaction can occur leading to covalent bonding. A complex comprises a central metal ion sharing a pair of electrons donated by another constituent, termed a ligand, acting as a Lewis base. The metal ion and ligand share a single water envelope. Ligands can be neutral (*e.g.* H_2O) or anionic species (*e.g.* Cl^-,HCO_3^-). A metal ion can co-ordinate with one or more ligands, which need not be the same chemical entity. Alternatively,the cation can share more than one electron pair with a given ligand thereby forming a ring structure. This type of complex is known as a chelate. They have enhanced stability, largely due to the entropy effect of releasing large numbers of molecules from the water envelopes.

Complex formation is an equilibrium process, and ignoring charges for the general case of a metal, M and ligand, L complex formation occurs as:

$$M + L = ML$$

for which the formation constant, K_1 is given by:

$$K_1 = \frac{\{ML\}}{\{M\}\ \{L\}}$$

A second ligand may then be coordinated as:

$$ML + L = ML_2$$

$$K_2 = \frac{\{ML_2\}}{\{ML\}\ \{L\}}$$

The equilibrium constant for ML_2 can be expressed solely in terms of the activities of the M and L:

$$\beta_2 = \frac{\{ML_2\}}{\{M\}\ \{L\}^2}$$

where β_2 is the product of K_1K_2 and is known as the stability constant. The case can be extended to include *n* ligands as:

$$\beta_n = \frac{\{ML_n\}}{\{M\}\ \{L\}^n}$$

The speciation of constituents in solution can be calculated if the individual ion activities and stability constants are known. This information is relatively well known with respect to the major constituents in seawater, but not for trace elements. There are some important confounding variables that create considerable difficulties in speciation modelling. Firstly, it is assumed that equilibrium is achieved, that is, neither biological interference nor kinetic effects prevent this state. Secondly, seawater contains appreciable amounts (at least in relation to the trace metals) of organic matter. However, the composition of the organic matrix, the number of available binding sites, and the appropriate stability constants are poorly known. Nonetheless, speciation models can include estimates of these parameters.Organic material can form chelates with relatively high stability constant and dramatically decrease the free ion activity of both necessary and toxic trace elements. Organisms may make use of such chemistry, producing compounds either to sequester metals in limited supply or to detoxify contaminants. Thirdly, surface adsorption onto colloids or suspended particles may remove them from solution. As with organic matter, an exact understanding of the complexation characteristics of the suspended particles is not available, but approximation can also be incorporated into speciation models.

Elements may be present in a variety of phases other than in true solution. Colloidal formation is particularly important for elements such as Fe and Mn which form amorphous oxyhydroxides with very great complexation characteristics. Adsorption processes cannot be ignored in biogeochemical cycling. Particles tend to have a much shorter residence time in the water column than do dissolved constituents. Scavenging of trace components by falling particles accelerates deposition to the sediment sink.

Several elements in seawater may undergo alkylation via either chemical or biological mechanisms. Type I mechanisms involve methyl radical or carbonium ion transfer and no formal change in the oxidation state of the acceptor element. The incoming methyl group may be derived for example from methylcobalamin coenzyme, *S*-adenosylmethionine, betaine, or iodomethane. Elements involved in Type I mechanisms include Pb, Tl, Se and Hg. Other reaction sequences involve the oxidation of the methylated element. The methyl source can be a carbanion from methylcobalamin coenzyme. Oxidative addition from iodomethane and enzymatic reactions have also been suggested. Some elements that can undergo such methylation processes are As, Sb, Ge, Sn, and S. Methylation can enhance the toxicity of some elements, especially for Pb and Hg. The environmental mobility can also be affected. Methylation in the surface waters can enhance volatility and so favour evasion from the sea, as observed for S, Se, and Hg. Methylation within the sediments may facilitate transfer back into overlying waters.

Element may exhibit multiple oxidation states in seawater. Redox processes can be modelled in an analogous manner to the ion pairing and complexation outlined previously. The information is often presented graphically in the form of a predominance area diagram, that is a plot of P*e* *versus* pH showing the major species present for the designated conditions. Although a single oxidation state might be anticipated from equilibrium considerations, there are several ways in which multiple oxidation state might arise. Biological activity can produce non-equilibrium species, as evident in the alkylated metals discussed above. Whereas Mn^{IV} and Cu^{II} might be expected, photochemical processes in the surface waters can lead to the formation of significant amounts of Mn^{II} and Cu^{I}.

SUSPENDED PARTICLES AND MARINE SEDIMENTS

Description of Sediments and Sedimentary Components

The sediments represent the major sink for material in the oceans. The main pathway to the sediments is the deposition of suspended particles. Such particles may be only in transit through the ocean from a continental origin or formed *in situ*. Sinking particles can scavenge material from solution. Accordingly, this section introduces the components found in marine sediments, but emphasizes processes that occur within the water column that lead to the formation and alteration of the deposited material.

Marine sediments cover the ocean floor to a thickness averaging 500 m. The deposition rates vary with topography. The rate may be several mm per year in nearshore shelf regions, but is only 0.2 to 7.5 mm per 1000 years on the abyssal plains. Oceanic crustal material is formed along spreading ridges and moves outwards eventually to be lost in subduction zones, the major trenches in the ocean. As a consequence of this continual movement, the sediments on the sea floor are no older than Jurassic, about 166 million years.

The formation of marine depends upon chemical, biological, geological, and physical influences. There are four distinct processes that can be readily identified. Firstly, the source of the material obviously is important. This is generally the basis on which sediment components are classified and will be considered below in more detail. Secondly, the material and its distribution on the ocean floor are influenced by its transportational history, both to and within the ocean. Thirdly, there is the deposition process which must include particle formation and alteration in the water column. Finally, the sediments may be altered after deposition, a process known as diagenesis. Of particular importance are reactions leading to changes in the redox state of the sediments.

Table 7.5 The four categories of marine sedimentary components with examples of mineral phases

Classification	*Mineral example*	*Chemical formula*
Lithogenous	Quartz	SiO_2
	Microcline	$KAlSi_3O_8$
	Kaolinite	$Al_4Si_4O_{10}(OH)_8$
	Montmorillonite	$Al_4Si_8O_{20}(OH)_4.nH_2O$
	Illite	$K_2Al_4(Si,Al)_8O_{20}(OH)_4$
	Chlorite	$(Mg,Fe^{2+})_{10}Al_2(Si, Al)_8O_{20}(OH, F)_{16}$
Hydrogenous	Fe-Mn minerals	$FeO(OH)—MnO_2$
	Carbonate fluoropatite	$Ca_5(PO_4)_{3-x}(CO_3)_xF_{1+x}$
	Barite	$BaSO_4$
	Pyrite	$Fe S_2$
	Argonite	$CaCO_3$
	Dolomite	$CaMg(CO_3)_2$
Biogenous	Calcite	$CaCO_3$
	Aragonite	$CaCo_3$
	Opaline silica	$SiO_2. nH_2O$
	Apatite	$Ca_5(F, Cl) (PO_4)_3$
	Barite	$BaSO_4$
	Organic matter	
Cosmogenous	Cosmic spherules	
	Meteoric dusts	

Source: R.M. Harrison, S.J. de Mora, S. Rapsomanikis, and W.R. Johnson, 'Introductory Chemistry for the Environmental Sciences', Cambridge University Press, Cambridge, 1991, 354 pp.

The components in marine sediments are classified according to origin. Examples are given in Table 7.5. Lithogenous (or terrigenous) material comes from the continents as a result of weathering processes. The relative contribution of lithogenous material to the sediments will depend upon proximity to the continent and the source strength of material derived elsewhere. The most important components in the lithogenous fraction are quartz and the clay minerals (kaolinite, illite, montmorillonite, and chlorite). The distribution of the clay minerals varies considerably. Illite and montmorillonite tend to be ubiquitous in nature, although the latter has a secondary origin associated with submarine volcanic activity. Kaolinite typifies intense weathering observed in tropical and desert conditions. Therefore, it is relatively enriched in equatorial regions. On the other hand, chlorite is indicative of the high latitude regimes where little chemical weathering occurs. The lithogenous components tend to be inert in the water column and represent detrital deposition. Nonetheless, the particle surfaces can be important sites for adsorption of organic material and trace elements.

Hydrogenous components, also known as chemogenous or halmeic material, are those formed within the water column. This may comprise primary material formed directly from seawater upon exceeding a given solubility product, termed authigenic precipitation. The best known example of authigenic material are ferromanganese nodules found throughout the oceans. Alternatively, secondary material may be formed as components of continental or volcanic origin become altered by low temperature reactions in seawater, a mechanism known as halmyrolysis. Halmyrolysis reactions occur in the estuarine environment, being essentially an extension of chemical weathering of lithogenous components. Such processes continue at the sediment—water interface. Accordingly, there are considerable overlaps between the terms weathering, halmyrolysis, and diagenesis.

Biogenous (or biotic) material is produced by the fixation of mineral phases by marine organisms. The most important phases are calcite and opaline silica, although aragonite and magnesian calcite are also deposited. Several plants and animals are involved, but the planktonic organisms are the most important with respect to the world ocean. The source strength depends upon the species composition and productivity of the overlying oceanic waters. For instance, siliceous oozes are found in polar latitudes (diatoms) and along the equator (radiolaria). The relative contribution of biogenous material to the sediments depends upon its dilution by material from other sources and the extent to which the material is dissolved in seawater. As noted previously, both calcareous and siliceous skeletons are subject to considerable dissolution in the water column and at the sediment—water interface.

Marine Pollution

Pollution comprises the perturbation of the natural state of the environment by anthropogenic activity. In the marine environment, the man-induced disturbances take many forms. Due to source strengths and pathways, the greatest effects tend to be in coastal regions. Such waters and sediments bear the brunt of industrial and sewage discharges, and are subject to dredging and spoil dumping.

Agricultural run-off may contain pesticide residuces and elevated nutrients, the latter of which may over-stimulate biological activity producing anoxic conditions. However, the deep sea has not escaped contamination. Although this has usually been in the form of crude oil, petroleum products, and plastic pollutants, the aeolian transport of heavy metals has enhanced natural elemental cycles, particularly with respect to lead.

Oil Slicks

Massive releases of oil have been caused by the grounding of tankers (*i.e. Torrey Canyon*, Southwest England. 1967; *Argo Merchant*, Nantucket Shoals, USA, 1976; *Amoco Cadiz*, Northwest France, 1978; *Exxon Valdiz*, Alaska, 1990) or by the accidental discharge from offshore oil platforms (*i.e. Chevron MP-41C*, Mississippi Delta, 1970; *Ixtox I*, Gulf of Mexico, 1979). The resulting oil slicks are essentially surface phenomena, with greatest impact where they impinge on coastal ecosystems. The slicks are affected by several transportation and transformation processes (Murray, 1982). With respect to transportation, the principal agent for the movement of slicks is the wind, but length scales are important. Whereas small (*i.e.* relative to the slick size) weather systems, such as thunderstorms, tend to disperse the slick, cyclonic systems can move the slick essentially intact. Advection of a slick is also affected by waves and currents. Oil slicks spread as a buoyant lens under the influence of gravitational forces, but generally separate into distinctive thick and thin regions. Such pancake formation is due to the fractionation of the components within the oil mixture. Diffusion can also act to transport the oil.

Several transformation mechanisms are operative. Evaporative loss of the more volatile components is a significant loss process, especially for light crude oil. Sedimentation can play a role in coastal waters. This occurs when rough seas bring dispersed oil droplets into contact with suspended particulate material, and the density of the resulting aggregate exceeds the specific density of seawater. Colloidal suspensions can comprise either water-in-oil or oil-in-water emulsions. These behave distinctly differently. Water-in-oil emulsification creates a thick, stable colloid which can persist at the surface for months. The volume of the slick increases and it aggregates into large lumps known as 'mousse', thereby acting to retard weathering. Conversely, oil-in-water emulsions comprise small droplets of oil in seawater. This aids dispersion and increases the surface area of the slick which can accelerate weathering processes, including microbial degradation.

Tributyltin

Tributyltin (TBT) is perhaps the most toxic constituent that has been knowingly introduced into the marine environment (Goldberg, 1986). TBT has been utilized as the active ingredient in marine anti-fouling paint formulations. Its potent toxicity and environmental persistence has resulted in a wide range of deleterious biological effects on non-target organisms. TBT is lethal to some shellfish at concentrations as low as 0.02 μg TBT-Sn l^{-1} Lower concentrations result in sub-lethal effects, such as poor growth rates and reduced recruitment leading to the decline of shellfisheries. The most obvious manifestations of TBT contamination have been shell deformation in Pacific oysters (*Crassosterea gigas*), and the development of imposex (*i.e.* the formation of male sex organs in females) in marine gastropods. This has caused dramatic population decline of gastropods at locations throughout the world. Laboratory experiments and field observations of deformed oysters and imposex indicate that adverse biological effects occur at concentrations below those detectable. Thus, a 'no effect' concentration has yet to be demonstrated. It is worth noting that TBT has been observed to accumulate in salmon, but it has not been shown to pose a public health risk.

TBT exists in solution as a large univalent cation and forms a neutral complex with C I. It is extremely surface active and so is readily adsorbed onto suspended particulate material. Such adsorption and deposition to the sediments limits its lifetime in the water column. Degradation, via photochemical reactions or microbially mediated pathways, obeys first order kinetics. Stepwise debutylation produces di- and mono-butyltin which are much less toxic in the marine environment than is TBT. As degradation rates in the water column are on the order of days to weeks, they are slow relative to sedimentation. TBT accumulates in the sediments where degradation rates are much slower, with the half-life being on the order of years (Stewart and de Mora 1990).

REFERENCES

E. Goldberg, M. Koide, J.S. Yang, and K.K. Bertine, 'Metal Speciation: Theory, Analysis and Application', ed. J.R. Kramer and H.E. Allen, Lewis Publishers, Chelsea, 1988, p.201.

E. Goldberg, Environment, 1986, 28, 17.

C. Stewart and S.J. de Mora, Environ. Techno., 1990, 11, 565.

S. Murray, 'Pollutant Transfer and Transport in the Sea', ed. G. Kullenberg, CRC Press, Boca Raton, 1982, Vol 2., p. 169.

CHAPTER 8

Soil And Soil Pollution

Introduction

Land, or more precisely its key component the soil, differs in many ways from the other parts of terrestial ecosystems, such as water and the atmosphere, in that it acts as a sink, a filter, and a bioreactor for contaminants.

It is the convention to use the term 'contaminated land' whether or not apparent harmful effects have been caused by the contaminants. This is in contrast to the widespread use of the term 'polluted' when substances of anthropogenic origin have caused obvious harm and 'contaminated' when they have not. Contaminated land is generally defined as 'land which represents an actual or potential hazard to health or the environment as a result of current or previous use'.

The composition of the 'soil' on contaminated land needs to be determined in order that the concentrations of contaminants comply with the critical ('trigger') concentrations stipulated for various end uses based on estimates of risk. The risk to materials and services include: sulphate attack of concrete, degradation of plastics and rubber by hydrocarbons, and fires or explosions from methane or hydrocarbons. The risks to humans include respiratory disorders and other illnesses from the inhalation of toxic fumes or dusts, direct ingestion of toxic compounds in contaminated drinking water or particles of soil, skin disorders from contact with chemicals, and consumption of food plants that have accumulated significant amounts of harmful substances.

In addition to contamination, anomalously high concentrations of cations and anions potentially harmful to either construction materials or crops, livestock, and humans, may be present in soils as a result of the weathering of some types of rocks. Examples include: high concentrations of SO_4^{2-} (which attacks concrete) from weathering sulphides in clay and shales, and heavy metals, such as Cd, Pb, and Zn, from weathering marine black shales or unexploited ore minerals. since these potentially harmful substances are not anthropogenic in origin, they should not be classed as pollutants or contaminants.

The phytotoxic effect of contaminants is of obvious economic importance where crops are being grown commercially but the greatest impact on human health is likely to occur where relatively bioavailable but potentially hazardous metals, such as Cd, are present in contaminated soil at sub-phytotoxic concentrations. Food crops on these soils may accumulate significant amounts of the metals without the contamination problem being recognized and this could have a chronic effect on the health of regular consumers of these crops.

It is important to recognize that although acute cases of land contamination attract a lot of attention and pose major risk, almost all the world's land surface is contaminated to some extent. The transport and deposition of atmospheric pollutants affects even very remote areas, although the extent of contamination is generally worst is the more industrialized northern hemisphere. Contaminants can be carried hundreds or thousands of kilometres by moving air masses.

Particles of soil blown thousands of kilometres from China caused a 'brown snow' event in the Canadian Arctic in April 1988. Around 4000 tonnes of soil was deposited over an area of 20 000 km^2 contributing appreciable quantities of organic micropollutants (PAHs and PCBs) in a very remote area. Radio-nuclides were dispersed over much of the earth's surface by the atmospheric testing of nuclear weapons in the 1950s and 1960s whilst large areas were affected by the accident at the Chernobyl nuclear reactor in the Ukraine in 1986 and an earlier accident at Windscale in England in 1957.

Apart from atmospheric deposition, the soil receives contaminants through direct placement. This can occur from the agricultural use of fertilizers and pesticides, the addition of large amounts of livestock manures and sewage sludges to agricultural land, and other types of waste disposal. However, direct deposition tends to give rise to localized contamination, for example a burnt-out dumped car on wasteland severely contaminates a few square metres, whereas the fallout from a metal smelter will have affected tend or hundreds of square kilometres, but with smaller quantities of contaminants.

Although almost all the soil on the earth's land surface is contaminated to some extent, it is the more acute cases of contamination which pose the most risk to living organisms and construction materials. It is therefore important to determine the extent of contamination, the contaminants involved, the risks they pose, and the most suitable way of either cleaning up the soil or managing it appropriately. This chapter will address these points briefly but the reader will need to refer to other more specialized books and papers for greater detail on the subject.

SOIL: ITS FORMATION, CONSTITUENTS, AND PROPERTIES

Soil is the complex biogeochemical material which forms at the interface between the earth's crust and the atmosphere and differs markedly in physical, chemical, and biological properties from the underlying weathering rock from which it has developed. Soil comprises a matrix of mineral particles and organic material, often bound together as aggregates, and populated by a wide range of micro-organisms, soil animals, and plant roots. The spaces between particles and aggregates form a system of pores which are filled with soil solution and gases. Since soil acts as a sink for contaminants and also a filter for water infiltrating to the water table, its physical, chemical, and biological properties are highly relevant to considerations of contaminated land and its reclamation.

Soil Formation

The soil profile (a vertical section from the surface to the underlying weathered rock) is the unit of study in pedology. Natural pedogenic processes bring about the differentiation of the soil material into distinct horizontal layers, called horizons, and the characteristics of these horizons forms the basis of soil morphological classification. A typical soil profile with a summary of the distribution of contaminants is shown in Fig. 8.1.

The type of soil which forms at any site is determined by the climate, the nature of the parent rock material on which it forms, the landscape (especially whether it is a free-draining or water collecting site), the vegetation and the time period over which the soil has been forming. These factors control the intensity and type of pedogenic processes operating at any site, and these processes determine the nature of the soil profile which develops. The major pedogenic processes include: leaching, elution, podzolization, calcification, ferralitization, laterization, salination, solodization, gleying, and peat formation. With the exception of the last two, most pedogenic processes involve the movement of material (usually the products of weathering) either down or up the soil profile, depending on the overall balance of precipitation and evaporation. Movement is predominantly downwards in humid regions and upward where evaporation exceeds precipitation but the extent will depend on the climate, the permeability of the soil and the site. Gleying is the creation of reducing conditions in soils, due to

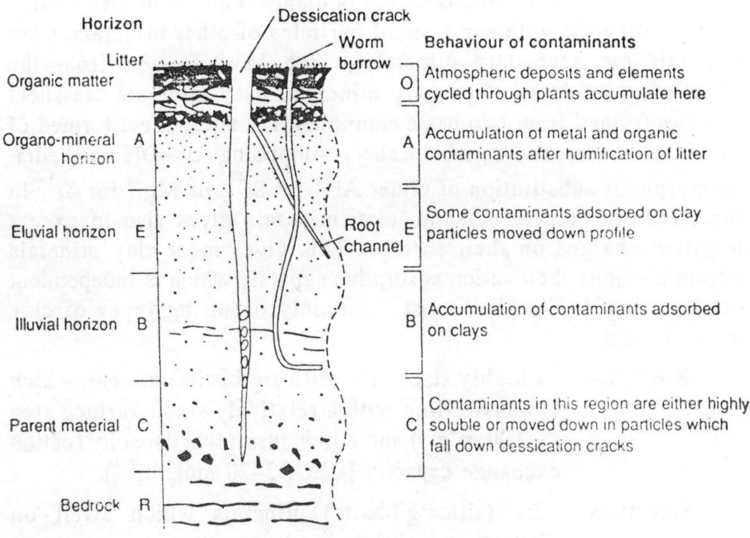

Fig. 8.1. Diagram of a soil profile showing horizon nomenclature and the general distribution of Contaminants.

intermittent or permanent waterlogging, and leads to the reduction and dissolution of hydrous oxides of Fe and Mn.

The pedogenic processes occurring within a soil will have a major influence on the behaviour of contaminants. For example, soluble or polar contaminants are more likely to be leached down the profiles of highly permeable soils and reach the water table than they are in impermeable clay soils or soils with relatively high organic matter contents in their uppermost horizon. Soil organic matter, especially colloidal humus, has a strong capacity to adsorb a wide range of substances. Soils in arid environments tend to have low contents of organic matter in their topsoils (<1 %) and are therefore less likely to adsorb organic nonpolar pollutants, such as PAHs and chlorinated hydrocarbons, than soils in humid temperate regions which have higher organic matter contents (1–10%). However, the relatively high free calcium carbonate content of arid soil increases their ability to adsorb many heavy metals.

SOIL CONSTITUENTS

Soil Minerals

The soil clay fraction is <2 μm in diameter and is formed mainly of clay minerals with some small particles of other minerals. Clay minerals are secondary minerals which have formed from the weathering products of primary minerals. Clay minerals are sheet silicates formed from two basic components: a silica sheet formed of Si—O tetrahedra, and a gibbsite sheet comprising Al—OH octahedra. Isomorphous substitution of either Al^{3+} for Si^{4+}, or Mg^{2+} for Al^{3+} in the structural sheets of some clay minerals gives rise to excess negative charges on their surface. This gives most clay minerals except kaolinite their cation adsorptive capacity which is independent of the soil pH. The four most commonly occurring types of clay minerals are:

Kaolinite –a highly stable 1:1 (silica:gibbsite) mineal, which is nonswelling with a relatively small surface area (5–100 m^2g^{-1}) and a low adsorptive capacity (cation exchange capacity [CEC] 2–20 mol_c kg^{-1}).

Smectites –2:1 (silica:gibbsite) minerals which swell on wetting and shrink on drying, they have a large surface area (700–800 m^2 g^{-1}) due to the access of soil solution to all lamellae surfaces which, together with isomorphous substitution, contributes to their relatively high adsorptive capacity (CEC 80–120 mol_c kg^{-1}).

Illite –a 2:1 clay like the smectites with a lower adsorptive and swelling/shrinking capacity and properties intermediate between kaolinite and smectites (surface area 100–200 $m^2 g^{-1}$ and CEC of 10–40 c $mol_c kg^{-1}$).

Vermiculite –a 2:1 mineral with a very high degree of isomorphous substitution of Mg^{2+} for Al^{3+} in the gibbsite sheet giving it a high CEC (100-150 c $mol_c kg^{-1}$) with an intermediate surface area (300–500 $m^2 g^{-1}$).

The other ubiquitous secondary minerals in soil are the hydrous oxides of Fe and Al which are the ultimate residual weathering products of soils; the anhydrous forms of the oxides are highly stable under the oxidizing conditions found in freely draining soil environments. The amount of these oxides present in a soil depends on the mineralogy of the parent rock, the degree of weathering and the oxidation/reduuction (redox) conditions. The ferromagnesian minerals such as olivine, augite, and biotite mica found in basic igneous rocks are the richest primary source of Fe. Poorly drained and waterlogged soils have reducing conditions which cause the dissolution of hydrous oxides. The main forms of Fe oxide are ferrihydrite ($Fe_2O_3 \cdot 2FeOOH \cdot nH_2O$), the freshly deposited form, and goethite (α–FeOOH). The Al oxides include amorphous $Al(OH)_3$ which slowly crystallizes to gibbsite (γ-$Al(OH)_3$). These hydrous oxides tend to occur as precipitates with the clay-sized fraction of soil minerals (<2 μm). The hydrous oxides have pH-dependent surface charges; in general they are positively charged under acid conditions and negatively charged under alkaline conditions.

Many other minerals may be found in soils, the most ubiquitous being resistant quartz grains which comprise much of the sand-size fraction of soils and are relatively unreactive. Other minerals tend to occur in fragments of weathering rock, either as clasts of parent material or erratics in fluvioglacial deposits.

Soil Organic Matter

The presence of organic matter and living organism distinguishes a soil from regolith (decomposed rock). All soils, in the pedological sense of the word, contain organic matter but the amount and type may vary considerably. The organic matter present in soils can be classified as either humic, or non humic material. Humic compounds are highly polymerized, colloidal, products of microbial decomposition of plant material, especially lignins. Non-humic material includes undecomposed or partially decomposed fragments of plant tissues and soil organisms. These latter include a wide range of species of bacteria, fungi, protozoa, actinomycetes, and algae, and many species of mesofaunna, such as earthworms, which fill an important ecological niche in the comminution of plant litter and its incorporation into the soil profile.

Although soil organic matter only tends to form a small percentage of the mass of the soil, it has a very great influence on soil chemical and physical properties especially with regard to the behaviour of contaminants. Humus has a high but pH-dependent CEC (<200 c $mol_c kg^{-1}$) and a strong adsorptive capacity for non-polar organic molecules. Soil micro-organisms can adapt to being able to degrade many different types of persistent contaminant molecules, such as chlorinated hydrocarbons and PAHs by the secretion of extracellular enzymes. However, the C—Cl bond does not occur in natural compounds and hence and chlorinated organic molecules are more difficult to degrade. Nevertheless, several species of micro-organism have been found to be able to degrade chlorinated compounds but there is normally a time lag while they adapt to the new molecules. During this period there is localized selection and multiplication of strains which have a mutation enabling them to break the bonds of the contaminants molecule and to tolerate the toxicity of either the contaminant or its intermediate decomposition products.

SOIL PROPERTIES

Soil Permeability

The voids between soil particles and aggregates form a continuous system of pores within the soil profile. Pores with a diameter greater then 30 μm tend to drain under gravity and are normally filled with air in dry weather. Pores smaller than 30 μm tend to retain water against gravity and much of this may be available to plant roots. Soluble contaminants infiltrating the soil profile will be subject to both interial drainage within the profile, diffusion between regions of different solute concentration, and, in most cases, adsorption onto the organo-mineral colloidal complex. In addition to the inter-particle pores, worm burrows, root channels, desiccation cracks, or excavations will lead to more rapid movement of contaminants down the profile, either in solution, or adsorbed on soil particles.

The permeability of soil can vary widely as a result of the nature of the soil constituents and, or, soil management; for example, clay soils have a relatively low permeability and sandy soils tend to be highly permeable. Compaction of soils by machinery, cultivations, or excavations when the soil is too wet, can reduce its permeability and may induce temporary reducing conditions in the upper part of the profile especially in soils with a high clay content.

Soil Chemical Properties. Soil pH

The pH of the soil is the most important physico-chemical parameter affecting plant growth and the behaviour of contaminants in soils. The soil pH is a measurement of the concentration of H^+ ions in the soil solution present in the pores of a soil which is in equilibrium with the negatively charged surfaces of the soil particles. Unless stated otherwise soil pH values are usually measured in distilled water but the use of dilute electrolytes (*e.g.* $CaCl_2$) more closely reflects the field situation.

Soil pHs are normally within the range 4–8.5 although the extreme range found over the world is pH 2–10.5. In general, soils in humid regions tend to have pHs between 5 and 7 and those in arid regions between 7 and 9. In temperate regions, such as the UK, the optimum pH for arable soils is 6.5 and 6.0 for grassland. Soil pHs can be raised by liming with $CaCO_3$.

In general, bacteria do not tolerate very acid conditions and so soils in which the microbial degradation of organic pollutants is desirable should be maintained at a pH of between 6 and 8. The mobility and bioavailability of most divalent metals are greatest under acid conditions and therefore liming is a way of reducing their bioavailability (except for the MoO_4^{2-} anion which is most available at high pH).

Redox conditions

The balance of oxidation-reduction conditions in soils mainly affects the species of elements such as C, N, O, S, Fe, and Mn although Ag, As, Cr, Cu, Hg, and Pb are also affected. The redox conditions in a soil are also a reflection of the oxygen supply for plant roots and soil micro-organisms. Redox equilibria are controlled by the aqueous free electron activity and can be expressed either as pe (negative log of electron activity) or *E* (millivolt difference in potential between a P_t electrode and H electrode). The conversion factor for the two expressions is: *E* (mV) = 5.2 pe. Micro-organisms catalyse redox reactions in soils and respiration by plant roots, soil fauna, and micro-organisms consumes a relatively large amount of oxygen. In situations of water logging, or exclusion of air by over-compaction, micro-organisms with anaerobic respiration predominate causing a change in the products of decomposition of organic matter (volatile fatty acids and ethylene, *etc.*) and in the speciation of susceptible metals.

The combined effects of redox conditions and pH on the behaviour of hydrous oxides in soils can be summarized by stating that either low pH or negative redox values result in the dissolution of hydrous oxides of Fe and Mn and, conversely small increases in either pH or redox values can lead to the precipitation of ferrihydrite or Mn hydrous oxides. Sulphate anions are reduced to sulphide below p*e* –2.0 and this can lead to the formation of insoluble sulphides of a wide range of metals. These sulphides tend to be insoluble and act as a temporary sink of the metal until redox conditions change and the sulfides are oxidized. The oxidation of sulphides, such as iron pyrites (FeS_2) results in a marked increase in the acidity of the soil.

SOURCES OF SOIL CONTAMINANTS

Soil receive contaminants from a wide range of sources, including:

(1) Atmospheric fallout from:
- fossil fuel combustion (oxides and acid radicals of S and N);
- Pb, PAHs, *etc.* from automobile exhausts;
- metal smelting operations (As, Cd, Cu, Cr, Ni, Pb, Sb, Tl, and Zn);
- chemical industries (organic micropollutants, Hg);
- waste disposal by incineration (TCDDs, TCDFs);
- radioisotopes from reactor accidents (*e.g.* Windscale, UK, 1957 and Chernobyl, USSR, 1986) and atmospheric testing of nuclear weapons;
- large fires (e.g. soot, PAHs, *etc.* from burning oil wells).

(2) Agricultural chemicals:
- herbicides (*e.g.* 2,4-D, 2,4,5-T containing TCDD, B, and As compounds);
- insecticides (chlorinated hydrocarbons, *e.g.* DDT, BHC);

- fungicides (Cu, Zn, Hg, and organic molecules);
- acaricides (*e.g.* 'Tar Oil');
- fertilizers (*e.g.* Cd and U impurities in phosphates).

(3) Waste Disposal (intentional/unintentional input to soil):

- farm manures (As and Cu in pig and poultry manures);
- sewage sludges (rich in heavy metals and organicpollutants – PAHs and PCBs, *etc.*);
- composts from domestic wastes (metals, *etc.*);
- mine wastes (coal mines–SO_4, *etc.*, metalliferous mines Cd, Cu, Pb, Zn, Ba, U, *etc.*);
- seepage of leachate from landfills;
- ash from fossil fuel combustion, incinerators, bonfires, and accidental fires;
- burial of diseased livestock on farmland.

(4) Incidental Accumulation of contaminants:

- corrosion of metal in contact with soil (*e.g.* Zn from galvanized metal, Cu and Pb from roofing, scrapyards, *etc.*);
- wood preservatives from fencing (PCP, creosote, As, and Cu);
- leakage from underground storage tanks (petrol, Chlorinated solvents); –warfare (organic pollutants from fuels, smoke, and fires, metals from munitions and vehicles);
- sports and leisure activities (Pb from gun shot and fishing weights, Pb, Cd, Ni, and Hg from discarded batteries, hydro-carbons from spilt petrol and lubricating oil).

(5) Derelict industrial sites–wide range of contaminants from production, waste disposal, and building demolition, *e.g.*:

- Gas works–phenols, tars, cyanides, As, Cd;
- Electrical industries–Cu, Pb, Zn, PCBs, solvents;
- Tanneries– Cr;
- Scrapyards–metals, PCBs, hydrocarbons.

A more comprehensive list of the most common contaminating uses of land includes: waste disposal sites, gas works, oil and petroleum refineries, and petrol stations, electricity generating stations, iron and steel works, non-ferrous metals processing, metal products fabrication and metal finishing, chemical works, glass-making and ceramics, textile plants, leather tanning works, timber and timber products treatment works, manufacture of integrated circuits and semiconductors, food processing, sewage works, asbestos works, docks and railway land, paper and printing works, heavy engineering installation, installations, processing radioactive materials, and burial of diseased farm livestock.

CHARACTERISTICS OF SOME MAJOR GROUPS OF SOIL CONTAMINANTS

Heavy Metals

Metals, such as Ag, Cd, Cr, Cu, Hg, Mn, Mo, Ni, Pb, Sb, Tl, U, V, and Zn tend to be strongly adsorbed by soil constituents, especially organic matter, 2:1 clays and hydrous oxides of Fe and Mn. Their mobility and bioavailability depends on the soil organic matter content, pH, and redox conditions. With the exception of Mo, metals are generally more mobile (bioavailable) under acid and reducing conditions and/or where the soil organic matter content is low (<2%). In addition to the relatively obvious sources of heavy metals, another major non-point source of Pb is game bird and clay pigeon shooting which can result in the dispersal of >1000 *t* Pb into the terrestrial environment annually in most technologically advanced countries.

The range of heavy metal concentrations found in sewage sludges ($\mu g\ g^{-1}$ in dry matter): Ag (<960), As (<30),Cd (<3410), Cr (< 40 600), Cu (50–8000) , Hg (< 55) , Mn (60 – 3900, Mo (< 40) , Ni (< 5300) , PB (29–3600), Sb (< 34). Se(< 10), Sn (40–700), V (20–400), Zn (91–49 000)[10].

Organic Contaminants

There are more than 20,000 organic contaminants known already and this number will increase as analytical methods are further refined and more studies made of materials containing wide ranges of organic pollutants, such as industrial wastes, sewage sludges, and landfill leachates.

Pesticides can be soil contaminants as a result of persistence after use on crops, run-off from treated land, accidental spillages, or pesticide manufacture. Although there are over ten thousand commercial pesticides formulations of around 450 compounds in use, they can be classified under relatively few groups of molecules. These compounds are used because of their inherent toxicity towards specific pests but there is a risk of the toxic properties affecting soil organisms, other beneficial plants and animals, and humans. This can occur by uptake through plants, through the food chain, or in contaminated drinking water which the pesticide, or its toxic decomposition product, reached either by leaching or in run-off.

Three main groups of insecticides are currently used in agriculture: organochlorines, organophosphates, and carbamates. The organochlorines have been widely used for up to 50 years and are the most persistent of all groups of pesticides. The persistence decreases in the order: DDT > dieldrin > lindane (BHC) > heptachlor > aldrin with half-lives of eleven years of DDT down to four years for aldrin. Organophosphates are highly toxic to humans and other mammals but are less persistent in the soil than organochlorines (six month half-lives for parathion, diazinon, and demeton). Carbamates are used to control a wide range of pests including molluscs, fungi, and insects but have a similar persistence to organophosphates. Aldicarb (or 'Temik') is a highly toxic carbamate that is used both as an insecticide and nematicide on potato and sugar beet crops. It is readily oxidized in the soil but its oxidation products are also highly toxic and readily leached and can cause ecological and human health problems. Some examples of the molecular structures of pesticides and other organic contaminants are shown in Fig 8.2.

There are six major groups of compounds used as herbicides: phenoxyacetic acids, toludines, triazines, phenylureas, bipyridyls, and glycines. The most important phenoxyacetic acids are 2,4-D and 2,4,5-T which have a persistence of up to eight months but are of particular environmental significance because they can be contaminated with dioxins (TCDDs). 'Agent Orange' the defoliant used by the US forces in Vietnam comprised a mixture of 2,4-D and 2,4,5-T and caused widespread contamination of soils by dioxin. The toluidines and triazines are fairly strongly adsorbed and have a persistence of up to twelve months although atrazine contamination of groundwaters is a serious problem in many intensive arable farming areas. The phenylureas tend to be fairly soluble and are rapidly leached. In contrast, the bipyridyls such as paraquat and diquat are cationic and strongly adsorbed on soil colloids $K_d$values of $<4.2 \times 10^4$ on montmorillonite) and are therefore very persistent. Fungicides comprise a more diverse group of compounds including inorganics, such as copper and mercury compounds, and a wide range of organic compounds.

Apart from pesticides, which are synthesized intentionally, a wide range of chlorinated compounds including dioxins (TCDDs) and dibenzodifurans (TCDFs) have been widel dispersed in the

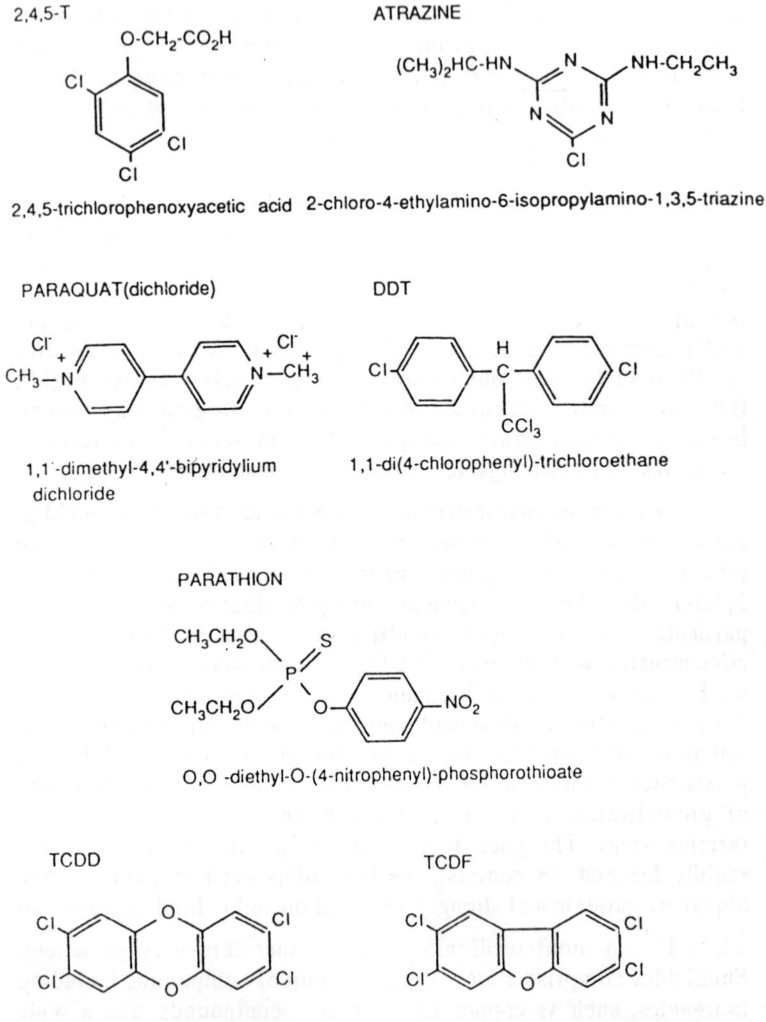

2,4,5-T

2,4,5-trichlorophenoxyacetic acid

ATRAZINE

2-chloro-4-ethylamino-6-isopropylamino-1,3,5-triazine

PARAQUAT(dichloride)

1,1'-dimethyl-4,4'-bipyridylium dichloride

DDT

1,1-di(4-chlorophenyl)-trichloroethane

PARATHION

O,O -diethyl-O-(4-nitrophenyl)-phosphorothioate

TCDD

2,3,7,8-tetrachlorodibenzodioxin

TCDF

2,3,7,8-tetrachlorodibenzofuran

Fig. 8.2 ***Examples of the molecular structures of pesticides and organic contaminants***

environment as a result of their accidental synthesis at relatively high temperatures, their highly stable structure, and their slow rate of degradation. Typical situations in which they are formed include: synthetic reactions where the temperature conditions become too hot (*e.g. production* of 2,4,5-T), the incineration of rubbish containing PVC and other sources o chlorides and aromatic compounds. Polychlorinated biphenyls (PCBs) are very stable and were manufactured for use in electrical transformers and capacitors and as plasticizers. They are found in soils around industrial and domestic waste tips and electronic component factories. Incineration temperatures need to be > 1200° C order to ensure the destruction of stable molecules like PCBs. Solvents used for degreasing electrical components such as trichloroethane are important atmospheric and groundwater contaminants and may also occur in soils around factories.

Sewage Sludge

Sewage sludges contain a wide range of environmental contaminants owing to the diverse sources of effluents discharged into sewers. This includes human excretion products, household chemicals, automobile fuels, lubricants and cleaning compounds, stormwater run-off from highways containing PAHs and other fuel combustion products, and effluents from many different industries. PAH concentrations in sewage sludges tend to range between 0.5 and 10 $\mu g\ g^{-1}$ in sludge dry matter with similar levels of PCBs although > 1000 $\mu g\ g^{-1}$ are sometimes found.

The organic contaminants frequently found in sewage sludges include:

- Halogenated aromatics (PCBs – polychlorinated biphenyls, PCTs– polychlorinated terphenyls., PCNs– polychlorinated naphthalenes, and polychlorobenzenes);
- Aromatic amines and nitrosamines;
- Halogenated aromatics containing oxygen and phenols;
- Polyaromatic and heteroaromatic hydrocarbons (PAHs);
- Halogenated aliphatics;
- Aliphatic and aromatic hydrocarbons;
- Phthalate esters;
- Pesticides.

Of all the organic contaminants listed here PAHs and PCBs are currently considered to constitute the greatest hazard to human health.

The range of heavy metal concentrations found in sewage sludges is given above.

A relatively high proportion of the sludge produced in many countries is applied to agricultural land as a means of disposal (67% in the UK). This sludge has many useful properties for agriculture (source of N and P and physical soil conditioner) but its use is limited by its concentrations of persistent contaminants. Sewage sludges are frequently used in the landscaping of derelict land where they act as a growth medium for plants grown for amenity purposes rather than food.

THE ADSORPTION OF CONTAMINANTS IN SOILS

Adsorption Reactions

Soils act as a sink for contaminants due to several adsorption which bind contaminants with varying strengths to the surfaces of the colloidal constituents of soils. This adsorption therefore delays or prevents the leaching of the contaminant down the soil profile to the water table, reduces its bioavailability to plants, and can affect the rate of decomposition of organic contaminants.

The concentration of solutes in the soil solution is determined by the interaction of adsorption and desorption processes. When contaminants are deposited on the soil surface they either react with the colloids in the soil aggregates at the surface or are washed into the soil profile in rain, irrigation water, or snow melt. Soluble contaminants will infiltrate the topsoil and enter the system of pores, whereas insoluble and hydrophobic organic molecules will bind to sites on the soil surface and become incorporated into the topsoil by movement of soil particles during cultivation, or excavations.

Several different adsorption reactions can occur between the surfaces of organic and mineral colloids and the contaminants. The extent to which the reactions occur will be determined by the composition of the soil (especially the clay mineral, hydrous oxide, and organic matter contents) and the soil pH, and the nature of the contaminants. The more strongly adsorbed contaminants are less likely to be leached down the soil profile and will tend to have relatively low biovailabilities. Ionic contaminants such as metals, inorganic anions, and certain organic molecules, such as the bipyridyl herbicides paraquat and Diquat are adsorbed onto surface charges on soil colloids. Non-ionic organic molecules, which includes most of the organic micropollutants and pesticides are adsorbed onto humic polymers by both chemical and physical adsorption mechanisms.

Adsorption of Cations and Anions in Soils

Ion exchange (or non-specific adsorption) refers to the exchange between the counter-ions balancing the surface charge on the soil colloids and the ions in the soil solution. In the case of cations, it is the negative charges on soil colloids which are responsible for cation exchange. The extent to which adsorbing soil constituents can act as cation exchangers is expressed as the cation exchange capacity (CEC) measured in c mol_c kg^{-1}. Some examples of the typical CEC values for soil colloidal constituents are

Soil organic matter	150–300 (c mol_c kg^{-1})
Kaolinite (clay)	2–5 (c mol_c kg^{-1})
Illite (clay)	15–40 (c mol_c kg^{-1})
Montmorillonite (clay)	80–100 (c mol_c kg^{-1})
Vermiculite (clay)	150 (c mol_c kg^{-1})
Hydrous oxides (Fe, Al, Mn)	4 (c mol_c kg^{-1})

Soil organic matter has a higher CEC at pH 7 than other soil colloids and therefore plays a very important part in all adsorption reactions in most soils even though it is normally present in much smaller amounts (1–10%) than clays (<80%). However, sandy soils with low contents of both organic matter and clay tend to have low adsorptive capacities and are a greater danger for contaminants infiltrating through the soil profile to the water table.

The negative charges on the surfaces of soil colloids are of two types:

(a) permanent charges resulting from the isomorphous substitution of a clay mineral constituent by an ion with a lower valency;

(b) the pH dependent charges on oxides of Fe, Al, Mn, and Si and organic colloids which are positive at pHs below their isoelectric points and negative above their isoelectric points.

Hydrous iron and aluminium oxides have relatively high isoelectric points (>pH 8) and so tend to be positively charged under most conditions whereas clay and organic colloids are predominantly negatively charged under alkaline conditions. With most colloids, increasing the soil pH, at least up to neutrality, tends to increase their CEC. Humic polymers in the soil organic matter fraction become negatively charged due to the dissociation of protons from carboxyl and phenolic groups.

The concept of cation exchange implies that ions will be exchanged between the soil solution and the zone affected by the charged colloid surfaces (double diffuse layer). The relative replacing power of any ion on the cation exchange complex will depend on its valency, it diameter in hydrated form and the type and concentration of other ions present in the soil solution. With the exception of H^+, which behaves like a trivalent ion, the higher the valency, the greater the degree of adsorption. Ions with a large hydrated radius have a lower replacing power than ions with smaller radii. For example, K^+ and Na^+ have the same valency but K^+ will replace Na^+, owing to the greater hydrated size of the Na^+ ion. The commonly quoted relative order of replaceability on the cation exchange complex of metal cations is:[8]

$$Li^+ = Na^+ > K^+ = NH_4^+ > Rb^+ > Cs^+ > Mg^{2+} > Ca^{2+} > Sr^{2+}$$

$$= Ba^{2+} > La^{3+} = H^+ (Al^{3+}) > Th^{4+}$$

Anion adsorption occurs when anions are attracted to positive charges on soil colloids. As stated above, hydrous oxides of Fe and Al are usually positively charged below pH 8 and so tend to be the main sites for anion exchange in soils. In general, most soils tend to have far smaller anion exchange capacities than action exchange capacities. Some anions, such as nitrates and chlroides are not absorbed to any marked extent but others, such as orthophosphates tend to be strongly adsorbed.

Some organic pesticides, such as the phenoxyalkanoic acid herbicides, such as 2,4-D, 2,4,5-T, and MCPA, exist as anions at normal soil pHs and are adsorbed to a limited extent by hydrous oxides and by H-bonding to humic polymers. However, this sorption is less marked than that which occurs with the cationic bipyridyl herbicides which are inactivated on contact with soil colloids.

Specific adsorption is a stronger form of adsorption, involving several heavy metal cations and most anions with surface ligands to form partly covalent bonds with lattice ligands on adsorbents, especially hydrous oxides of Fe, Mn, and Al. This adsorption is strongly pH specific and the metals and anions which are most able to form hydroxy complexes are adsorbed to the greatest extent. The order for the increasing strength of specific adsorption of selected heavy metals is:

$$Cd > Ni > Co > Zn >> Cu > Pb > Hg$$

Adsorption and Decomposition of Organic Contaminants

The adsorption of non-ionic and non-polar pesticides and other organic contaminants occurs mostly on soil humic material. Since the highest content of soil organic matter in soils occurs in the surface horizons of soils, there is a tendency for most organic contaminants to be concentrated in the topsoil. Migration of organic contaminants down the profiles of soils occurs to the greatest extent in highly permeable sandy or gravelly soils with low organic matter contents. High concentrations of water soluble soil organic matter can cause enhanced mobility and leaching of organic contaminants in soils due to the binding of the contaminants to the soluble ligand. Soils will vary in their contents of water soluble soil organic matter, but applications of sewage sludge, animal manure, and compost results in increased concentrations.

In the case of pesticide contaminants, most are relatively insoluble and do not move down the soil profile but exceptions to this are the organophosphate insecticides, phenoxyacetic acid herbicides, and bipyridylium herbicides which can be leached. In general, most pesticide and other organic contaminants that reach water courses from soils have been washed into the water adsorbed on soil particles in run-off and not leached down through the profile (see Fig 8.2 for structures of some pesticide and other organic contaminants molecules).

In some cases where organic wastes containing organic contaminants, such as sewage sluges and some composts, are applied to soils, they act as both a source and a sink of contaminants (*e.g.*, PAHs, PCBs, and heavy metals). A range of distribution coefficients are found for the same contaminant molecules in different soils and this is considered to be largely due to variations in the soil organic matter content. For example, k^{oc} (distribution coefficient for organic carbon) values ranging from 540-1730 have been reported for Lindane and 240– 1290 for naphthalene. Positive correlations are frequently found between the concentrations of pesticide residues and the soil organic matter content.

Adsorption of organic contaminants depends on their surface charge and their aqueous solubility, both of which are affected by the soil pH. Adsorption of non-polar organic contaminants onto soil organic matter will not occur in the presence of oils. Microbially synthesized surfactants can help to accelerate the rate of degradation of hydrocarbon oils in contaminated soils. For many organic contaminants, adsorption onto soil colloids and the presence of water are important factors promoting decomposition by micro-organisms. Ross, lists the types of degradation of organic molecules as: (*a*) Non-biological degradation which includes : hydrolysis, oxidation, reduction, and photodecomposition; and (*b*) Microbial decomposition, often involving specially adapted micro-organisms. This type of decomposition normally follows a first order (exponential) type of reaction after an initial lag period while the micro-organisms become adapted to the substrate. In most cases, non-biological degradation processes such as photodecomposition and volatilization can occur at the same time as microbially catalysed reactions.

The range of factors affecting the degradation of organic contaminants by micro-organisms include: soil pH, temperature, supply of oxygen and nutrients, the structure of the contaminant molecules, their toxicity and that of their intermediate decomposition products, the water solubility of the contaminant, and its adsorption to the soil matrix (and therefore the organic matter content of the soil). Temperature effects tend to result in adsorption decreasing with increased temperature. Most pesticide adsorption reactions are exothermic. Volatilization losses tend to be greatest at high temperatures.

The persistence of organic contaminants in soils is determined by the balance between adsorption onto soil colloids, uptake by plants, and transformation or degradation processes (determined by the amount of contaminant, its form, and soil properties). In the case of pesticides, organochlorine molecules are regarded as being highly persistent with a duration in the soil of 2-5 years but some estimates put this up to 10 or more years.

Brief details of commonly used analytical methods used for detection of soil contaminants are summarized in Table 8.1. Samples of soil are normally air-dried or over dried <30°C but care should be taken over the possibility of harmful vapours being released from the samples.

Table 8.1 Summary of the analytical methods used for soil contaminants

Contaminant	*Method*
Heavy metals (in acid digests or partial extracts)	Flame atomic absorption spectrophotometry (FAAS) or inductively coupled plasma-atomic emission spectrometry (ICP-AES)
As, Bi, Hg, Se, Sn and Te (in acid digests or partial extracts)	Hydride generation atomic absorption spectroscopy (HGAAS) using sodium borobydride in NaOH
Borate (water soluble)	ICP-AES (using quartz or plastic apparatus instead of glass)
Organic pollutants (in organic solvents)	Gas chromatography (GC) or combined with Mass Spectrometry (GC-MS)
Cyanides	Colorimetrically–pyridine pyrazalone (blue) reaction
Sulphates (water soluble)	Elution though cation exchange resin followed by titration of eluate with standard NaOH
Sulphates (total)	Dissolution in HCl precipitation of Al and Fe followed by gravimetric determination using $BaCl_2$
Chlorides (in HNO_3)	Volhard's Method, back titration with ammonium thiocyanate after initial precipitation with $AgNO_3$
pH (in distilled water or $CaCl_2$ or KCl)	Electrometrically using glass electrode

THE EFFECTS OF SOIL CONTAMINATION

Soil contamination can restrict the options available for the use of land because of the potential hazards posted by the contaminants. Some examples of the hazards are:

	Hazard	**Examples of Contaminants**
(1)	Direct ingestion of contaminated soil (mainly children or animals)	As, Cd, Pb, CN^-, coal tars phenols
(2)	Inhalation of dusts and vapours from contaminated soil	toluene, benzene, xylene, various organic solvents Rn, Hg, metal-rich particles
(3)	Uptake by plants of contaminants hazardous to animals and people through food chain	As, Cd, Pb, Tl, PAHs
(4)	Phytotoxicity	SO_4^{2-}, Cu, Ni, Zn, CH_4
(5)	Deterioration of building materials and services	SO_4^{2-} SO_3^{2-}, Cl^-, coal tar, phenol, mineral oils ,organic solvents
(6)	Fires and explosions	CH_4, S, coal dust, oils, petroleum, tar, rubber, high calorific organic wastes (old landfills)
(7)	Contact of people with contaminants during demolition	coal tar, phenols, asbestos, radionuclides, PAHs TCDDS, *etc.*
(8)	Contamination of water	CN^-, SO_4^{2-} ,soluble metals, solvents, pesticides

Contaminants in soil can affect humans by absorption into the body through oral, inhalation, or cutaneous pathways and their effects are usually related to the amount absorbed into the body. Inhalation is an important route of intake for volatile compounds such as organic solvents and mercurial compounds. Dust particles < 7 μm can penetrate alveoli in the lungs and are a hazard to workers involved in moving contaminated soil and possibly to people using playing fields constructed on contaminated land. Oral intake is a particularly serious problem with young children playing on contaminated land when they lick their fingers. Oral intake in adults is mainly through the consumption of soil on vegetables (and also the contaminants accumulated in the plants). Cutaneous absorption is only a problemwith lipophyllic organic solvent.

CRITICAL CONCENTRATIONS FOR CONTAMINANTS IN SOILS

Having established that an area of soil is contaminated, it is necessary to make a decision about the action that needs to be taken in order to avoid unnecessary risk of health effects or of damage to structures. For this purpose, various sets of critical concentrations are in use around the world. The ranges of critical concentrations used in two European countries (UK and The Netherlands) for the interpretation of contaminated land analytical data are given as examples in Tables 8.2.

In the Netherlands, a system has been used which involves three indicative values: A–the 'normal' reference value, B–the test value to determine the need for further investigations, and C–the value above which the soil definitely needs cleaning up. Examples of these indicative values for three typical contaminants are (in $\mu g\ g^{-1}$): for Cd, A = 1, B = 5, and C = 20; for total complex cyanides, A = 5, B = 50, and C = 500; and for total PCBs, A = 0.05, B = 1.0, and C = 10.

Table 8.2. UK Department of the Environment Trigger Concentrations for Environmental Contaminants17,19 (total concentrations except where indicated)

Contaminant	*Proposed uses*	*Threshold (Trigger Concentrations* $\mu g\ g^{-1}$*)*	*Action*
5 Contaminants which may pose hazards to health			+
As	Gardens, allotments	10	
	parks, playing fields, open space	40	
Cd	Gardens, allotments	3	
	parks, playing fields, open space	15	
Cr (hexavalent*)	Gardens, allotments	25	
	parks, playing fields, open space	1000	
Cr	Gardens, allotments	600	
	parks, playing fields, open space	1000	
Pb	Gardens, allotments	500	
	parks, playing fields, open space	2000	
Hg	Gardens, allotments	1	
	parks, playing fields, open space	20	
Se	Gardens, allotments	3	
	parks, playing fields, open space	6	
(*hexavalent Cr extracted by 0.1 M HCl adjusted to pH 1 at 37.5°C)			
Contaminants which are phytotoxic but not normally hazardous to health			
B (water soluble) .	Any uses where plants grown	3	
Cu (total)	Any uses where plants grown	130	
(extractable**)		50	
Ni (total)	Any uses where plants grown	70	
(extractable)		20	
Zn (total)	Any uses where plants grown	300	
extractable)		130	

**Extracted in 0.05 M EDTA

+Action concentration yet to be specified.

REFERENCES

B.J. Alloway, 'heavy Metals in soils', ed., B.J. Alloway, Blackie, Glasgow, 1990.

G.W. Brummer, 'The Importance of Chemical Speciation in Environmental Processes', Springer Verlag, Berlin, 1986.

I. Kogel-Knabner, P. Knabner, and H. Deschauer, 'Contaminated Soil '90', ed. F. Arendt, M. Hinsenveld, and W.J. van den Brink, Kluwer Academic, Dordrecht, 1990, p. 323.

H. Kishi and Y. Hasimoto, 'Contaminated Soil '90', ed. F. Arendt, M. Hinsenveld, and W.J. van den Brink, Kluwer Academic, Dordrecht, 1990, p.331.

M.J. Beckett and D.L. Sims, 'Contaminated Land', ed. J.W. Assink and W. J. van den Brink, Martinus Nijhoff, Dordrecht, 1986, p.285.

Interdepartmental Committee on the Redevelopment of Contaminated Land, 'Guidance on the Assessment and Redevelopment of Contaminated Land', Guidance Note 59/83, Department of the Environment, London, 1987.

W.L. Lindsay, 'Chemical Equilibria in soils', John Wiley, Chichester, 1979.

D. Sauerbeck, 'Scientific Basis for Soil Protection in Europe' ed. H. Barth and P. L'Hermite, Elsevier, Amsterdam, 1987, p. 181.

S. Ross, 'Soil Process', Routledge, London, 1989.

R.f. White, 'Introduction to the Principles and Practice of Soil Science', 2nd Edn. Blackwells, Oxford, 1987.

N. Brady, 'The nature and properties of Soils', 10th Edn., Macmillan, New York, 1990.

M.J. Singer and D.N. Munns, 'Soils: An Introduction', Macmillan, New York, 1987.

F.A.M. de Haan., 'Scientific Basis for Soil Protection in Europe', ed. H. Barth and P.L'Hermite, Elsevier. Amsterdam, 1987, p. 211.

E.M. Bridges, 'World Soils', 2nd Edn., Cambridge University Press, Cambridge, 1978.

CHAPTER 9

Effects of Pollutants

Introduction

Pollution is 'the introduction by man into the environment of substances liable to cause hazards to human health, harm to living resources and ecological systems, damage to structure or amenity, or interference with legitimate use of the environment'.

There are, on the one hand, a group of substances which we recognize as pollutants and, on the other, organisms on which pollutants exert their effect. A basic maxim of toxicology is that all substances administered at sufficiently high doses are harmful to biota. Pollutants, by definition, are first introduced into the environment and then dispersed along pathways during which they may undergo transformations to more innocuous forms or to forms that are potentially more hazardous. One important aspect which influences the impact of a pollutant is the amount discharged into the environment. Conceivably, an organism may be able to cope with a small amount of a very toxic substance, whereas an otherwise essential one may reach overwhelming proportions and adversely affect many organisms. Pollutants encompass a broad range of chemical and physical properties and these properties strongly influence both the behaviour of pollutants in the environment and the manner in which they exert their effect on biological systems.

Man is responsible for releasing vast quantities of many different chemical substances into the environment each year, the majority of these anthropogenic substances are waste products generated by industry and society consuming the manufactured goods. Man-made fabrics and fibres, pharmaceuticals, fertilizers, pesticides, paints, and building materials, as well as chemicals for industrial processes, are just some of the products of the chemical industry that are integral to almost every aspect of modern living. Many such substances are natural constituents of the environment, others are man-made synthetic chemicals.

Inevitably wastes generated during manufacture, the substances themselves, and perhaps their degradation products, are released into the environment. Some such as dioxin and methyl cyanide, are released as a result of sudden accidents; others such as sulphur dioxide, carbon dioxide, and oxides of nitrogen from fossil fuel combustion, lead from petrol-driven engines, and mercury from chlor-alkali plants, are continuously discharged, while still others such as pesticides are intentionally released by man. In addition, hazardous chemical wastes are buried either on land or at sea.

Pollutants exert their effect on individuals, but recognition of an effect is only seen when large numbers and even whole communities are affected. The rate of supply or the amount of pollutant reaching a *receptor* (organism, population, or community) is referred to as the exposure and an effect is a biological change caused by an *exposure.* Directly toxic substances must gain access to the exterior membranes or interior components of tissues and cells in order to exert their effect. The amount that is taken into an organism is referred to as the *dose*; essentially it is a function of concentration and the period of exposure, although dose is more correctly defined as the amount of substance received at the site of effect. In many situations, however, this is impracticable since the site of action is not known.

A pollutant may be supplied in one large package (*acute exposure*) ; alternatively, an equivalent amount may be supplied at a lower concentration over an extended period (*chronic exposure*). Similarly, adverse effects or damage induced by a pollutant are frequently referred to as being acute or chronic. Acute effects are observed almost immediately following exposure ; they are rarely reversible and are very often fatal, Chronic effects or damage follow a period of prolonged exposure. Various biochemical and physiological disturbances may become apparent which are usually sub-lethal effects. They may be outwardly manifested as visible or clinical symptoms of damage, *e.g.* chlorotic regions on plant leaves, and inability to maintain homeostatic balance, co-ordinate activities, or breed. These symptoms are often quite diffuse and not specific to a particular pollutant. Following cessation of exposure, chronic damage is frequently reversible, although continued exposure may prove fatal. However, it may be argued that any severe disturbance will impair the efficiency of an individual and so shorten its life.

As well as the direct impact of pollutants on living species, other secondary effects require careful consideration. These relate to community and habitat changes that may follow initial pollution damage. The obvious examples include disturbance of predator-prey relationships following a dramatic decline of one or more species in a foodweb ; reduced turnover in biogeochemical cycling of major elements if, say, decomposer organisms are affected. These secondary impacts are a feature of the ecosystem level of organization.

Pollution studies are very wide ranging and so extend over a multitude of situations involving interactions of pollutants with living organisms. Damage to certain groups of organisms arouses more concern than others. Clearly man and his resources species—livestock, crops, and fisheries, for obvious social and economic reasons are two groups which engender most concern. Surveying the range of effects of pollutants, it is evident that these also can be arranged on a scale of increasing concern, with cellular disruption arousing less interest than a situation proving acutely toxic.

Exposure

Exposure is the amount of pollutant reaching a receptor, and it is commonly expressed as the concentration of pollutant in the media (*e.g.* air, water, soil) and in food that is available to the receptor (individual, population, community). It is important to take into account the period of exposure to estimate the rate of supply to a receptor and in turn the intake (or dose). In the case of the more persistent pollutants, the amount or concentration in an organism or some part of an organism is frequently a function of exposure. Consequently such measurements can be used to indicate exposure.

Exposure may arise from a single source of supply, as in the transfer of phytotoxic gases from the atmosphere to the leaf surfaces of plants. Alternatively more than one pathway may contribute to the total exposure, for example man's exposure to lead comes from water, air, and food.

Routine measurements of exposure are not always representative of the actual amount that reaches a receptor. In the aquatic environment the concentration of pollutants vary with season, time of day, magnitude of freshwater run-off, depth of sampling, intermittent flow of industrial effluent, and hydrological factors such as tides and currents, Man's exposure to atmospheric pollutants such as Pb, SO_2, and smoke can be particularly difficult to assess. Many samples are required to iron out variation due to methods of collection and analysis as well as actual spatial and time differences. Very often assessments are based on data from a small number perhaps only one monitoring station. The size distribution of particles containing lead (Pb) can vary considerably between different locations. It has been shown that, in London (UK) 60% of particles were less than 0.3 μm and only 1% were above 10 μm, whereas in Los Angeles (USA) only 30% were less than 0.3 μm and a greater proportion were above 10 μm. This is of considerable importance for estimating lead exposure to man by inhalation. This issue is further confounded by the fact that man is a very mobile animal. Many adults spend a part of their day at work and children are at school and consequently such groups may experience different exposure regimes during such times. We all spend extended periods indoors removed from the outside air and certain susceptible groups such as the aged would probably spend most of their time inside. Furthermore, such habits as tobacco smoking, which is a very direct form of air pollution, quite obviously increases a person's exposure to an array of substances and it is well known that cigarette smoking provides the major supply of cadmium (Cd) to habitual smokers ; smoking 20 cigarettes can result in a daily intake of 2—4 μg Cd.

Much of the research on the effects of gaseous air pollutants on plants has been done in the controlled environment of growth chambers. The concentration of the gas in question in the inlet or outlet of the chambers is frequently equated with plant exposure. These values however, can be quite different from the actual exposure at the leaf surface. Design and size of growth chamber, density of plants, and air velocity are just three of many plant and environmental factors that can influence the concentration gradient between ambient air around the plants and that in contact with the leaf surfaces. The crux of the matter is the effect of these and other factors on the thickness of the boundary layer of air that surrounds the leaves. Gases must pass through this almost laminar flow of air by the relatively slow process of molecular diffusion to gain access to the leaf surfaces. In situations where varied responses had been reported for seemingly for seemingly equivalent exposures, it is now apparent that at least a part of the discrepancy can be explained by the different ways in which exposure was determined.

Whenever an element such as a metal, as opposed to a molecular, species, is considered as a pollutant, the form in which it is exposed to a receptor influences its bioavailability and hence its potency. As a rule the free ion (*e.g.* Cd^{2+} , Cu^{2+} , Pb^{2+}) is the most bioavailable and therefore the most toxic, although organometallic compounds such as methylmercury and tributyl tin are more completely absorbed than their free ion counterparts. In soils and aquatic systems, metals are partitioned between solid and liquid phases and within each, further partitioning or speciation occurs among specific ligands, determined by ligand concentration and the strength of each metal—ligand association. Consequently at any one time, the amount of free-ion available for uptake by organisms is less, and in many instances much less, than the total concentration.

Absorption

Absorption is the process whereby a substance traverses the body membranes. By and large, lipid soluble substances are absorbed more efficiently than polar water soluble substances. Lipid soluble substances diffuse across the lipid bilayer of membranes and those that are more lipid soluble (*i.e.* those with a larger octanol-water partition coefficient) tend to be absorbed more efficiently. Membrane proteins serve to transport polar substances such as metals, although certain forms which are more lipid soluble or of reduced polarity traverse membranes by passive diffusion and this is the most likely explanation for the essentially complete absorption of methylmercury.

The characteristics of the interface between the environment and an organism strongly influence absorption of a substance. In mammals the main routes of entry for environmental chemicals are through the lungs, the skin, and the gastrointestinal tract. The physico-chemcial properties of a pollutant and the nature of exposure strongly influence the amount presented at each portal of entry. In terrestrial mammals, including man, penetration through the skin is an unimportant route of entry for environmental pollutants. It is a membrane which is relatively impermeable to aqueous solutions and most ions, although many of the man-made organic chemicals, because of their lipid solubility, can penetrate the dermal barrier. Such situations tend to be confined to occupational exposures.

Atmospheric pollutants occur as gases or as particulate matter. The site and extent of absorption of inhaled gases are for the most part, determined by their water solubility. For instance, sulphur dioxide (SO_2) is a soluble gas, which is mainly absorbed in the upper respiratory tract whereas a less soluble gas such as nitrogen dioxide (NO_2) reached the lower airways. Inhalation is the most important route of uptake for chemical mercury vapour, and the major site of absorption is alveolar tissue where about 80% of inhaled mercury vapour is absorbed.

The fraction of inhaled particulate material deposited in the various parts of the respiratory tract is a function of particle size, and the fraction absorbed from the tract is dependent on the chemical nature of the aerosol. Only particles less than 2 μm diameter penetrate as far as the alveolar region. Larger particles are trapped in the upper tracheobronchial and nasopharyngeal regions where they may be either absorbed or transported up the pharynx entrained in mucus propelled by ciliary action. Subsequently, these particles are swallowed and become available for absorption in the gastro-intestinal tract. Particles taken up from the alveoli (the most important site of absorption) may pass directly into the bloodstream or be retained in lung tissue. For respirable particles of lead, at a particle size of 0.05 μm about 40% may be retained but for larger sizes, *e.g. 0.5* μm only about 20% is deposited and probably only about 50% of the deposited lead is absorbed into the blood, although this will depend on the solubility of lead compounds in the particles. Deposition in the alveolar region of inhaled cadmium (Cd) aerosols up to 2 μm diameter is 20-35% of which less than 50% is absorbed; the rest is exhaled or swallowed.

In humans, absorption through the gastro-intestinal tract is an important route of entry for many environmental chemicals. The circulatory system is closely associated with the intestinal tract, and once toxicants have crossed the epithelium, entry into capillaries is rapidly effected. Veinous blood flow from the stomach and intestine introduces absorbed materials to the hepatic portal vein, resulting in transport to the liver, which is the main site of metabolism of foreign compounds. Exposure to metals in the general environment is usually greater via food and drink than via air. However, the absorption of ingested lead and cadmium in adults is normally relatively low, being about 10% for lead and 5% for cadmium. Although absorption is influenced by various dietary factors and increased absorption of lead has been found in cases of low dietary calcium. Absorption of dietary lead is much higher in young children. In the case of inorganic mercury compounds absorption from foods is about 7% of the ingested dose ; in contrast, gastro-intestinal absorption of methylmercury is practically complete.

The gills of aquatic animals present another type of interface between the external medium and an organism and are an important route of entry for water-dispersed pollutants. They present a large surface area for diffusion, and at the same time continual circulation of water across the gill filaments ensures maximum exposure.

One other type of external surface membrane worthy of note is that of the leaves of plants. Leaf surfaces are covered by a waxy cuticular layer, interspersed with stomatal pores which may occur on both leaf surfaces or on a single surface (usually the lower). Stomata, as well as regulating uptake of carbon dioxide and water loss, are the major route of entry for gaseous pollutants. Rate of uptake of gaseous pollutants into a leaf is a function of several physical factors ; one such factor is the resistance to diffusion caused by the boundary layer which in turn depends on the velocity and turbulence of airflow over the surface, and another factor is the stomatal resistance. Before a pollutant can gain access and cause injury within a plant cell it must first enter into solution in the extracellular water enveloping the cell wall. In solution, SO_2 is active in the form of either HSO_3^- or SO_3^{2-}. Ozone (O_3) is less soluble, but since it is a highly reactive molecule, it is thought that decomposition products such as hydroxyl radicals and other free radicals are important reactive species produced from reactions involving organic compounds.

Changes in stomatal aperture induced by air pollutants has attracted considerable attention. This is not surprising since any change in gaseous flux to and from metabolic sites in mesophyll tissues may eventually affect overall growth and yield of plants. The effect of SO_2 on stomatal aperture is very complex, initial studies found enhanced stomatal opening in *Vicia faba* plants exposed to SO_2. It was also shown that relative humidity strongly influenced the direction of the response ; humidities above 40% enhance stomatal apertures, whereas at humidities below 40% apertures decreased. A host of different species have now been investigated and it would appear that there is no uniformity of response between species ; both concentration of pollutant gas and duration of exposure influence the outcome.

Pathways Of Pollutants

Once inside an organism, a pollutant may follow a number of different pathways. In general terms four main routes can be identified (Fig 9.1) some molecules are metabolized (converted) into other compounds which are frequently less toxic to the organism than the parent compound, although some are more toxic. A second pathway involves storage in certain issues, *e.g.* lead in bone, cadmium in kidney, and DDT in tissue with a high fat content. Thirdly, a pollutant and its metabolites may be excreted from an organism. Because metabolites are often more easily excreted than the original pollutant, metabolism is frequently an essential preliminary to excretion. The remaining fraction is available to exert an effect at the site of action. The fundamental effect is a biochemical event, for example lead inhibits the action of ALA-D, an enzyme in

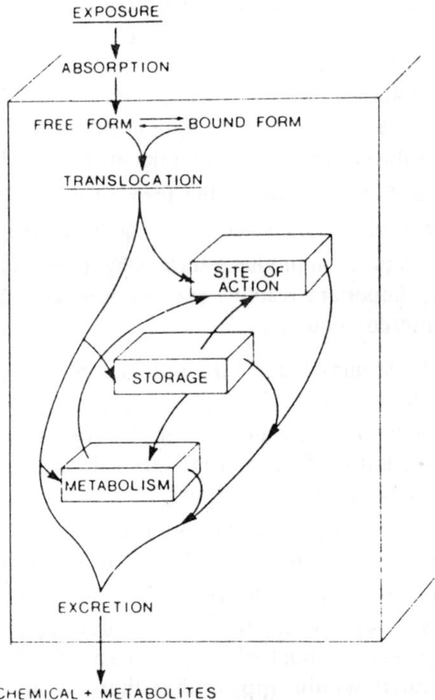

Fig.9.1 *Diagrammatic representation of the possible internal pathways followed by pollutants* (Adapted from T.A.Loomis, 'Essentials of Toxicology', 2nd Edn., Lea and Febiger, Philadelphia, 1974)

the haem synthesis pathway, although the significance of this to the overall toxicity of lead is not clear. Perhaps surprisingly, the primary toxic lesion of many pollutants is unknown. At low exposures, it is envisaged that most of the pollutant would be stored and/or metabolized and excreted and little if any would be available to have toxic effect. As exposure is increased these controlling processes are progressively overwhelmed, primary biochemical lesions appear which in turn leads to major physiological damage and ultimately death of the organism.

Internal pathways of the major phytotoxic gases are short. They exert their effect near the point of entry. For mesophyll tissues of plant leaves, for example, much of the damage due to sulphur dioxide involves disruption of chloroplasts and depression of photosynthesis and also changes in stomatal aperture. It is widely believed that much of the effect to ozone involves permeability changes in the membranes of palisade cells. Pollutants like ozone are so transitory that it is virtually impossible to detect them within tissues ; their presence is detected from the effects induced. While exposure to sulphur dioxide can increase plant sulphur content three-to four-fold, it is rapidly incorporated into organic molecules, notably the amino acids, glutathione, and cysteine. In fact the balance between incoming sulphur dioxide and its destruction is probably a critical factor in resistance to acute injury in different varieties and species of plants. Nitrogen dioxide is similarly incorporated into amino acids such as glutamine and asparagine.

For may metals, storage or accumulation within a tissue or an organ is essentially a detoxication step. Inorganic lead is transported in the bloodstream attached to red blood cells (erythrocytes) but a major proportion, about 90% in adults and 70% in children, accumulates in bone. The biological half-life in this tissue is about 10 years but turnover of lead in the bloodstream and soft tissues is rapid and responds quite quickly to changes in lead intake and exposure.

In normal healthy humans about 50% of the cadmium in the body is in the kidneys and liver with about one-third in the kidneys alone. The biological half-life in these tissues is probably more than 10 years and as a consequence cadmium accumulates with age. A significant proportion occurs bound to metallothioneins, which are low molecular weight proteins rich in sulfydryl groups. In this bounds form cadmium is less available to exert a toxic action. A number of other trace metals, including zinc, copper, mercury, and silver also induce and bind to metallothioneins. The precise role of metallothioneins is unclear and the regulation of intracellular concentrations of metals also involves other mechanisms, for example metals may be incorporated into extracellular structures such as carbonate granules, or bound to intracellular components such as nuclei mitochondria, lysosomes, and phosphate granules.

DDT, PCBs, and other chlorinated organic compounds are concentrated in fat deposits of a wide range of fauna, including birds and mammals. Tissues with the largest amount of fat accumulate the highest concentrations of these lipid-soluble compounds. In normal circumstances there is slow turnover of these compounds in the body's fat reserves. However, during periods of stress and starvation, mobilization of the fat deposits can release the stored organochlorines into the bloodstream which in turn may induce toxic effect and ultimately death. Laboratory studies have shown that starved animals will sometimes die of DDT poisoning whereas those adequately fed appear unharmed. The large number of guillemot deaths recorded in September 1969 from the coasts around the Irish Sea may have been due to a combination of stress and mobilization of organochlorine residues, perhaps PCBs, into the blood stream.

A feature of most, if not all, vertebrate and invertebrate animals is a built-in capacity to biotransform a wide range of foreign compounds to less toxic entities that are more easily eliminated from the organism. In higher organisms biotransformation mainly takes place in hepatic tissues and it essentially consists of two phases ; phase 1 is a mixed-function oxygenase system (MFO) which oxidizes foreign compounds by such mechanisms as hydroxylation, dealkylation, and epoxidation; phase two involves conjugation reactions by which glucoronide and sulphate derivatives are formed from the oxidized products of phase one. The net effect is to convert lipophilic foreign compounds to water-soluble metabolites which facilitates their elimination from the animal. Numerous hydrocarbon compounds, including various phenolic and benzene compounds, are known to undergo biotransformation to water-soluble conjugates. The influence of biotransformation in regulating tissue concentrations and hence the toxicity of a substance can be deduced from studies which have used inhibitors of the biotransformation pathways. Salicylamide is a potent inhibitor of the phase II glucuronide conjugation reactions and it has been shown that pretreatment of fish with salicylamide significantly increases the toxicity of phenol as a consequence of inhibition of the formation of phenylglucuronide. In the case of the organophosphorus insecticide parathion, the metabolite paxon is the actively toxic product but pretreatment of fish with sesmex (an inhibitor of MFO reactions) prevents the oxidation of parathion to the active paxon with the result that its toxicity was much reduced, In animal tissues DDT is rapidly converted to the more persistent DDE and this explains the fact that DDE, rather than DDT is the most widespread and abundant form in wildlife.

A fundamental relationship in pollution and ecotoxicology studies in that which relates exposure, body burden, or accumulation, and toxicity of a substance. Residues in an organism (or some part of it) form an important link between exposures and biological response or damage. For the most persistent pollutants, such as metals and organochlorine compounds, enhanced exposure from food or the surrounding media generally results in a greater concentration in an organism. When uptake (absorption) of a substance exceeds elimination (including metabolism) accumulation occurs in the whole organism or some part of it. Eventually a steady state may be attained when uptake and elimination are equal and at this stage the amount in the organism is proportional to the average daily intake and inversely proportional to elimination rate.

Assessment Of Effects

In practice, three methods of approach are employed in assessments of ecological and health effects of pollutants :

(1) standardized laboratory-based experiments ;

(2) field surveys and epidemiological studies ;

(3) field experiments and microcosm or mesocosm studies.

(1) Experimental studies may be concerned with uptake, accumulation, bio-transformation, and the toxicity of a substance or mixtures of substances. Acute toxicity tests are widely used and they usually form the basis of assessing the potential hazard of chemicals. The LD_{50} *(or* LC_{50})—the dose (or concentration) which is lethal to 50% of a test population over a set period of time (24,48, or more usually 96 hours)—is the parameter most commonly derived in such tests. These simple tests which are carried out under standardized conditions provide a useful means of screening chemicals and ascertaining their relative toxicity. However alteration of the experimental conditions can substantially modify toxicity values. All to frequently acute toxicity tests are based on a few well-known easily reared test species and, as a result marked differences in sensitivity of wildlife species are overlooked. Many incidents of poisoning in birds by carbamates and organophosphorus compounds could have been avoided had LD_{50} data extended to

more than one species. In one incident a number of geese (*Anser anser* and *Anser fabalis brachyrhynchus*) died from ingestion of lethal doses of carbophenothion, which is one of the organophosphorus insecticides that has replaced dieldrin as a seed dressing protecting winter wheat from wheat bulb fly. Original registration data, based on toxicity to chickens indicated a low avian oral LD_{50} ; however it is now known that the species of geese are very much more sensitive to carbophenothion than other bird species particularly chickens.

Chronic toxicity tests by their nature are undertaken at lower exposures for more prolonged periods and therefore they provide a more realistic indication of toxicity. Response criteria that have been used to indicate chronic stress include growth, feeding rate, respiration, reproductive activity and success behaviour, and bioaccumulation. An integrated physiological response, that has been used in chronic toxicity studies in the laboratory and the field, is scope for growth using the common mussel, *Mytilus edulis.* A detailed description of the methodology of measuring scope for growth may be found in Widdows and Johnson (1988). Basically, the index measures the energy available for activity, growth, and reproduction from assimilation of food, after respiratory and excretory requirements have been satisfied. In the presence of a pollutant this energy may decrease in response to a decreased feeding rate and expending metabolic energy to deal with stress. With regard to the ecological effects of offshore oil production in the North Sea, there is good agreement between scope for growth response in mussels exposed to elevated polycyclic aromatic hydrocarbons and community responses such as changes in species diversity.

(2) Field surveys and epidemiological studies, in basic terms, relate pollutant exposure to a biological response. Much of the remainder of this chapter examines various aspects of this relationship. However some general points are worth noting. Field evidence relating a pollutant to an adverse response does not constitute a cause—effect relationship, but merely demonstrates a statistical correlation. There are many practical problems concerned with estimating the true exposure of pollution to a group of orgnisms. Equally it is often difficult to obtain a true or representative indication of size of population, because of uneven distributions and fluctuations over time. The

impact of pollution on wildlife depends as much, perhaps more, on the duration of damage as on immediate losses.

(3) Field experiments, in many ways bridge the important gap between purely experimental methods and field surveys. This type of approach has proved particularly valuable in the study of the effects of gaseous air pollutants on plants. Basically plant growth chambers are set up in the field, one set receives unfiltered air containing air pollutants at ambient concentrations and the other set receives air from which the pollutants have been removed by filtration. A number of studies, some of which are referred to in subsequent paragraphs, have used this approach to good effect. It is important to realize that this is still fundamentally a correlative response, since effects are related to the pollutants which the operator chooses to measure.

Increasingly, microcosms and mesocosms are used to examine the fate and effect of chemicals in the environment. Such systems include artificial ponds and streams which are colonized with natural assemblages of plants and animals and then subjected to different degrees of chemical stress. In other situations a part of the lake, estuary, or the sea may be segregated using enclosures such as large rubber tubes and the entrapped communities exposed to various combinations of pollutants.

ECOLOGICAL AND HEALTH EFFECTS OF AIR POLLUTANTS

Health Effects

Health effects of air pollution in the general population are associated with an increase in mortality and worsening of heart and lung conditions during episodes of dense smogs with in which certains air pollutants accumulate to high concentrations. A number of incidents have been reported in different parts of the industrialized world, for example in the Meuse Valley in 1930 and in Donora in 1948. The most notorious incident occurred in London (UK) in early December 1952 when 3500—4000 deaths above the norm were recorded during an exceptional smog episode that lasted, unremittingly, for 5 days. At the time, much of the air pollution was due to coal combustion, with numerous near-ground level sources generating considerable quantities of smoke and suphfur dioxide. The fog developed across the capital as a consequence of a very stable high pressure zone and an inversion layer. This pervented the dispersal of pollutants and concentrations built up to very high levels, the particles of smoke acted as condensation nuclei which added further to the density of the fog. The epidemic went largely unnoticed by the population of London and it was only when the death certificates for the whole of the London area were later examined that the sudden upsurge in the number of deaths became apparent. Deaths were mainly confined to the elderly and those people with a history of heart and lung diseases. The central areas of London, where the fog was at its densest and most persistent, showed the greatest increases in mortality; some 200% more than the average for that time of year. Estimates indicate that sulphur dioxide and smoke concentrations (48 h mean) during the London episode attained very high values and were in the region of 3.7 mg m^{-3} (1.3 ppm.) for SO_2 and above 4.5 mg m^{-3} for smoke.

It is generally accepted that the combined effected of sulphur dioxide and smoke particles on the respiratory tract were responsible for the excess deaths and for exacerbating heart and lung disease in susceptible people. To the healthy majority of the populace of London the pollutant ladened fog was merely an inconvenience.

Follow-up studies in other large cities concentrated on more moderate day-to-day variations of mortality and morbidity in relation to pollution levels, A WHO Task Group provided a summary of the salient features of many of these studies and the collated data have formed the basis for developing short-and long-term exposure-effect relationships. Short-term exposures of 500 $\mu g\, m^{-3}$ (24 h mean) for both sulphur dioxide and smoke can be expected to result in excess mortality among the elderly and the chronically sick. Exposures of 250 $\mu g\, m^{-3}$ (24 h mean) to both pollutants are likely to lead to worsening of the condition of patients with existing respiratory disease. With regard to long-term exposure, increased prevalence of respiratory symptoms among both adults and children and increase frequency of acute respiratory illnesses in children are more likely to occur when annual mean concentrations of sulphur dioxide and smoke exceed 100 $\mu g\, m^{-3}$. WHO advocate, as a guideline, that 24 h mean values of sulphur dioxide and smoke should remain below 100-150 $\mu g\, m^{-3}$, and with an annual mean below 40-60 $\mu g\, m^{-3}$.

The relationship between photochemical smog episodes and human health are less clear cut. Early epidemiological studies, based mainly in Los Angeles, failed to establish direct relationships between increased mortality rates and the frequency of smog episodes. Although eye irritations (perhaps due to peroxyacyl nitrate compounds) and upper respiratory tract discomforture and common complaints during smog episodes. Increased breathing difficulties in heavy smokers and asthmatics and reduced performance in people indulging in physical activity are associated with periods of high oxidant concentrations. Laboratory and field studies of adults who exercise heavily for short periods of time have provided evidence for the existence of short-term reversible decrements in pulmonary function to ozone concentrations at or near the USA National Ambient Air Quality Standard of 0.12 p.p.m. In recent years there are indications, in several parts of the world, that high ozone levels are associated with an increased risk of asthmatic attack. There is evidence that the prevalence of asthma may be increasing in a number of countries, including the United States and it is also possible that asthma is becoming more severe. Recent population-based studies have shown that respiratory function may be reduced proportionately to increasing concentrations of ambient O_3.

Ecological Effects

The effect of air pollution on terrestrial ecosystems such as forests is dependent on the nature of the pollutants and the magnitude of exposure. At relatively low exposures ecosystems absorb or act as a sink to the influx of pollutants and little or no harm can be detected, in certain circumstances such as inputs of nitrogen compounds into a nitrogen deficient forest increased growth may even be observed. At intermediate or moderate exposure, physiological damage occurs, growth may be reduced, and sensitivity to disease may increase. In grossly polluted situations, widespread destruction of vegetation and denudation of soils occurs.

Immediately surrounding heavily industrialized areas and, most notably, in the vicinity of large point sources such as metal smelting operations, fumes rich in sulphur dioxide, and metals have caused extensive damage to vegetation. Well known examples are found around large smelters like those as Sudbury and Wawa, Ontario, and at Copper Hill, Tennessee, where hundreds to thousands of square kilometers of landscape have been denuded. Smaller examples are found in the Nangatuck Valley in Connecticut, around a zinc smelter at Palmerton, Pennsylvania, and the Lower Swansea Valley in the UK. Many of these activities have either closed down or emission controls have been installed and so the gross disturbances observed previously around such sources are increasingly uncommon. However severe damage to forests still occurs at very polluted sites in Europe, notably in Czechoslovakia, East Germany, and Poland and the nature and extent of this damage is only recently coming to light.

Extensive areas of Europe and North America are subject to moderately elevated levels of gaseous air pollution, for example ozone has caused widespread damage to forests in California. A point worth emphasizing is that a polluted air mass rarely contains a single phytotoxic agent. In recent years it has been increasingly realized that regional scale air pollution is concerned with complex gas—aerosol mixtures that may include varying proportions of SO_2, nitrogen oxides (NO_x) ammonia (NH_3), ozone (O_3) and acid aerosols.

Vegetation injury caused by photochemical smog was first reported in the Los Angeles basin in 1944 and it has continued to be a chronic problem in southern California for more than 40 years. It was soon established that O_3 was the main phytotoxic agent in the smog complex. Ozone has caused widespread injury to agronomic and horticultural crops and natural and managed forest ecosystems, not only in California but also in many other states where metrological conditions and primary pollution concentrations were favourable. Ozone is the most economically damaging air pollutant to vegetation in the USA. Large scale injury to tobacco crops in the eastern USA has been reported over a number of years. Extensive pine needle damage to Ponderosa and Jeffrey pines in western locations and white pine in the east were due to atmospheric oxidants. In the mixed conifer forest ecosystems in the mountains of Southern California, the dominant tree species, Ponderosa and Jeffrey pines, for 30 years or so, suffered annual mortalities of about 3% which means that in this period hundreds of thousands of trees have died. On one selected plot, monitored over a 20 year period, the standing volume of timber had decreased by 28%. Other field studies have indicated that Ponderosa pine stands have generally undergone significant reduction of height growth, radial growth, and total wood volume.

Evidence that ambient ozone conentrations in Britain during summer can inflict deleterious effects on vegetation comes from, a series of experiments, using open-top chambers, at a rural site near Ascot in Berkshire, some 32 km west of London. At this location SO_2 and NO_x concentrations are generally low, but episodes of high ozone concentrations have been regularly recorded. During three such incidents between 1978 and 1983, *Pisum sativum, Trifolium repens*, and *T. pratense* grown in open-top chambers and receiving unfiltered air, developed visible leaf necrosis typical of O_3 damage. Concentrations of O_3 during the course of these experiments exceeded 0.1 p.p.m., which is considered to be the threshold above which visible leaf necrosis appear in these sensitive plants. Field observations in the surrounding areas revealed symptoms typical of O_3 injury in a variety of crop plants. It has since been realized that *Pisum sativum* crops grown in the UK for a number of years, frequently developed necrotic lesions typical of ozone injury without the cause being known.

In recent years there has been considerable concern over the decline in forest health in western Europe. The problem first came to light in the late 1970s and early 1980s in West Germany when Norway spruce (*Picea abies*), the dominant silviculture tree in mid-Europe, began to show increasing signs of defoliation and needle discolouration. Extensive death of the fine root system has also been reported. These symptoms have since been observed in the UK, France, the Benelux countries, Scandinavia, the Alps (Switzerland and Austria), Czechoslovakia, and Poland. The decline in forest health is not confined to Norway spruce, other conifer and broad leaved species have been affected. It has been argued that whilst regional declines of individual species can occur, for example due to disease epidemics, the synchronous decline of several species implicates a common cause such as air pollution. In 1987 and 1988 a survey of the severity of damage across the European Community (EC) showed that the forests of southern Europe are in better health than those of central and northern Europe. In the major coniferous forests of northern Europe, 15%, 22%, and 28% of *Pices abies, Pices sitchensis*, and *Abies alba* trees, respectively, are unhealthy. Among the hardwoods of northern and central Europe, 16%, 15%, and 12% of *Quercus robur, Quercus petrea*, and *Fagus sylvatica*, respectively, show symptoms of damage.

Forest decline is frequently associated with foliar deficiency of certain nutrients, and in particular magnesium deficiency, *e.g.* at elevated sites in West Germany, although in others areas potassium deficiency has been found. However, the cause of the nutrient deficiency is unknown. There are reports of increased foliar leaching of mineral nutrients due to acidic deposition or other pollutants but this is thought to be of secondary importance and the supply of mineral nutrients from, the soil is thought to be more critical. There is growing evidence that the deposition of sulphur and nitrogen pollutants has significantly modified soil chemistry and plant nutrition. Increased nitrates and sulfates in the soil solution can increase the leaching of cations such as calcium and magnesium from soils and enhance soil acidification. Soil solution chemistry, notably decreases in calcium and/or magnesium to aluminium ratio, is known to affect root development and hence water and nutrient uptake.

In addition to effects on photosynthesis and stomatal conductance, there is also evidence that certain gaseous air pollutants acting singly and in combination (SO_2 , SO_2/NO_2 , and O_3) have an effect on the relative distribution of growth above and below ground, with root growth being more severely affected than shoot growth. This may be a consequence of interference with phloem translocation and assimilation in the plant. Other effects of air pollutants include damage to epicuticular wax layer of leaves and needles which can lead to increased water loss and also there are reports that air pollutants can delay the onset of winter hardening. It is well known that plants in polluted areas are more susceptible to attack buy insect pests and it has been shown in experiments that exposure of plants to small or medium doses of SO_2 and/or NO_2 results in increases of the population growth of aphids feeding on the plants. This in turn can result in significant increases in pest damage to plants.

Therefore the cause of forest decline is likely to involve a number of factors, including air pollution and environmental factors interacting in some way, with one set predisposing forest stands to attack and damages from another set.

Ecological Effects Of Acid Deposition

In recent decades, numerous streams, rivers, and lakes in various regions of western Europe and north-eastern North America have become progressively more acidic. It is now generally accepted that acid deposition with major acidic, or acidifying ions is the cause of fresh water acidification in geologically sensitive areas throughout western Europe and north-eastern North America. Geologically sensitive areas are those with slowly weathering granites and gneiss rocks that support acidic and weakly buffered soils. Freshwater acidification is typified by a loss of acid neutralizing capacity (essentially the carbonate buffering system), decrease in pH by as much as 1—2pH units, increases in sulphate, nitrates, ammonium ions, aqueous aluminium, and metals such as manganese and zinc. Acidification is responsible for the loss and depletion of fish populations from numerous freshwater ecosystems in parts Norway, Sweden, UK, Canada, and USA.

Freshwater organisms generally maintain their internal salt concentration by active uptake of ions (in particular Na^{+} and Cl) from water against a concentration gradient. Substantial experimental evidence and field data indicate that fish mortality at low environmental pHs is primarily caused by a failure to regulate internal salt concentrations at gill surfaces. A characteristic symptom of acid-stressed fish is therefore a reduction in the salt or electrolytic ion concentration of blood plasma.

At an early stage it was realized that toxicity is not simply a function of acidity. Field data indicated that fish mortality occurred in waters of pH 5, whereas laboratory experiments with purely acid solutions failed to reproduce these results and significant mortalities only occurred at pH 4. It has since been shown that an important influence on the toxicity of acidified rivers and lakes is the dissolved concentration and chemical species of aluminium. Aluminium is more soluble in acidic environments and various surveys of lakes and rivers have revealed that concentrations of aqueous aluminium increases in acidified waters. Because of reactions with hydroxide radicals, aluminium toxicity to fish is pH dependent and the most bioavailable forms of aluminium occur at pH 5. Several reports have shown the effects of the interaction of acidity and soluble aluminium on fish survival. The survival of brown trout fry is reduced in concentrations of aluminium of 250 $\mu g\, L^{-1}$ and greater, especially if calcium is low. The growth rate of brown trout is reduced at a concentration greater than 20 $\mu g\, L^{-1}$, a concentration more than an order of magnitude lower than the total aluminium typical of any acid waters. Moreover the most toxic combinations were found with elevated aluminium concentrations at pH 5.

EFFECTS OF METAL POLLUTION IN HUMANS

Three forms of mercury are important in the environmental cycles of the element, they are elemental mercury (Hg^{o}), divalent mercury (Hg^{2+}), and methylmercury ($CH_3 Hg^{+}$) . Methylmercury is a particularly toxic species of mercury and it is easily absorbed across the external membranes of animals.

Several major episodes of mercury poisoning in the general population have been caused by consumption of methyl-and ethyl-mercury compounds. Methylmercury is a neurotoxin and the main clinical symptoms of poisoning reflect damage to the nervous system. The sensory, visual, and auditory functions, together with those of the brain areas, especially the cerebellum, concerned with co-ordination, are the most common functions to be affected. Symptoms of poisoning progressively increase in severity in line with increased exposure, as follows : (1) initial effects are non-specific symptoms which include paraesthesia, malaise, and blurred vision (2) in more severe cases, concentric constriction of the visual field, ataxia dysarthria, and deafness appear more frequently ; and (3) in the worst affected cases, patients may go into a coma and die. The effects in severe cases are irreversible due to destruction of neuronal cells. There is a latent period, usually of several months between the onset of exposure and the development of symptoms.

The Iraqi outbreak

This epidemic of methylmercury poisoning occurred in agricultural communities in Iraq in the winter of 1971-72. Over 6000 people were admitted to hospitals in provinces throughout the country and over 400 people died in hospital with methylmercury poisoning. The poisonings stemmed from mis-use of imported high-grade seed grain treated with alkylmercury fungicide. The imported seed was intended for sowing but in many areas the grain was ground directly into flour and used in the daily baking of homemade bread.

The Minamata and Niigata Outbreaks

In Japan, two major epidemics of methylmercury poisoning have occurred, one in the Minamata Bay area and the other in Niigata. In each case the problem arose as a result of the local population consuming seafood contaminated with methylmercury. Mercury compounds, including methylmercury were released from industrial sources into the aquatic environment and subsequently this resulted in the accumulation of methylmercury in seafood. These outbreaks of poisoning were first discovered during the 1950s and the early 1960s and, by the mid 1970s, about 1000 cases (with 3000 suspected cases) in the Minamata area and over 600 in the Niigata area had been recorded. By and large symptoms of poisoning followed a very similar pattern to the Iraqi epidemic, although an important difference was in the nature of the exposure which was lower but more prolonged.

Because mercury even from a natural sources in fish is predominantly in the methylmercury form, there is concern that certain groups of people who depend largely on a fish diet will be exposed to slightly elevated intakes throughout their lives and hence possibly accumulate mercury to toxic levels. One such group is the Canadian Indians but because of several confounding factors, not least of which are high incidences of malnutrition and alcoholism, the assessment of health risk due to methylmercury is difficult to ascertain. One study, involving 35000 samples obtained from 350 communities, found that over two-thirds had mercury blood concentrations within normal limits ($<20 \mu g L^{-1}$ but 2.5% (over 900 individuals) had levels in excess of 100 $\mu g L^{-1}$ and which therefore could be considered as a group 'at risk' and therefore in need of close surveillance.

Mercury analysis of segments of hair can provide a meaningful index of past exposures and hence body burdens. As a guideline, blood mercury concentrations of 200-500 $\mu g L^{-1}$, hair concentrations of 50—125 $\mu g\, g^{-1}$, and a long-term intake of 3—7 $\mu g\, kg^{-1}$ body weight are likely to be associated with the onset of the initial symptoms of methylmercury poisoning, such as paraesthesia.

Clinical and epidemiological evidence indicates that prenatal life is more sensitive than adult life to the toxic effects of methylmercury. The first indications came from Minamata in the early stages of the outbreak, where it was found that mothers who were only slightly poisoned gave birth to infants with severe cerebral palsy. A similar situation has been reported in the Iraqi outbreak of 1971-72 with infants which had been prenatally exposed showing severe damage to the central nervous system. Using the maximum maternal hair concentration during pregnancy as an index, the lowest level at which severe effects have been observed was 404 $\mu g\, g^{-1}$.

Lead is a neurotoxin and the overt toxic effects of lead have been known for many centuries. Probably the first reported cases of lead poisoning due to environmental sources was a group of children diagnosed as having lead palsy by clinicians at the Brisbane Hospital in Queensland, Australia at the turn of the century. A total of ten cases of lead poisoning were found by Health Officials and it was later shown that the source of the lead was lead-based paint which was turning to powder on the walls of homes and railings.

Lead poisoning due to exposure to lead-based paint has affected a large number of children over the years. Most of the epidemiological studies have been centered on the United States of America. The disease is confined to children, especially those living in inner city areas in dilapidated buildings with surfaces of flaking and peeling lead-based paint. Children playing in the vicinity can take in particles and flakes of paint by inhalation and hand-to-mouth activities. Certain children have a craving for eating non-food items such as flaking paint. This habit called 'pica' and the more normal hand-to-mouth activities of children are capable of introducing excessive amounts of lead into the body, *e.g.* a square centimetre of paint may contain over one milligram of lead.

ECOLOGICAL IMPACT OF ORGANOCHLORINE COMPOUNDS

A wealth of literature exists on the relationship between DDT (and its metabolite DDE) and the phenomenon of eggshell thinning in various bird populations. It was first discovered in the peregrine falcon (*Falco peregrinus*) in the UK. This particular falcon is widespread throughout Eurasia and North America, feeding almost entirely on live birds caught in flight. From about 1955 onwards, for no obvious reason, numbers of falcons rapidly declined in southern England and subsequently this decline spread northwards to parts of the Scottish highlands. By 1962, in the UK, some 51% of all known pre-war territories had been deserted and this figure was as high as 93% for southern England. At the time biologists in several countries were also investigating populations of birds that had slowly declined to critical levels. It emerged that the species all had residues of DDT, its metabolites, other chlorinated hydrocarbon insecticides, polychlorinated biphenyls (PCBs), and other chemicals in their tissues.

It is well accepted that DDE is the primary cause of eggshell thinning in many kinds of birds and some bird species are more susceptible than others to DDE-induced shell thinning. Eggshell thinning was definitely a major cause of low reproductive success and population decline in some species, but chlorinated hydrocarbons probably contributed to declines in other ways. In general terms, reproductive trouble tends to increase as shells become thinner and thinning of over 20% is likely to result in reproductive failure and population decline.

The toxicity of PCBs to wildlife is quite variable, exposures as low as 0.1 $\mu g\ l^{-1}$ are toxic to certain fish species, while others can survive concentrations of over 1000 $\mu g\ l^{-1}$. Experimental studies with mink and ferret have established that PCBs are highly disruptive of reproductive processes. Mason reports the results of an experiment in which American mink (*Mustela vison*) were fed diets of PCBs. Females with intakes of 3.3 mg PCB per kg food for 66 days, produced fewer young and of those that were born the survival rate was very low. At higher intakes no young were produced at all. Tissue concentration of 50 mg kg^{-1}PCB was associated with the onset of reproductive failure. The otter which is a closely related species to the mink has declined markedly over wide areas of Europe since the 1950s. In particular animals from those populations showing the greatest decline tend to have PCB concentrations in their tissues of over 50 mg kg^{-1}.

In recent decades there has been a rapid decrease of seal populations in the Baltic and in the Dutch Wadden sea. Between 1950 and 1970 the population in the Wadden sea dropped from more than 3000 to less than 500 animals. The reduction in numbers is associated with low breeding success. The populations from both the Baltic and Wadden were found to contain high levels of organochlorine compounds and in particular PCBs.

Effects related to contamination of PCBs have been found in the seal populations, they include hormonal imbalances and immunosuppression. In the light of recent large scale seal deaths, which have been found to be due to a virus related to the canine distemper virus, there is some concern that such outbreaks may be related to a breakdown in their immune systems.

REFERENCES

F.Moriarty, 'Ecotoxicology : The Study of Pollutants in Ecosystems', 2nd Edn., Academic Press, London, 1988.

E-D. Schulze, and P.H. Freer-Smith,'Acid Deposition—Ist Nature and Impacts', eds. F.T.Last and B. Watling, Proc. Roy. Soc. Edinburgh, 1991, Section B,97,p.155.

P.R.Miller, and J.R.McBride, 'Air Pollution and Forest Decline', eds. J.B. Bucher and I.Bucher-Wallin, 1989, p.61.

J.N.B.Bell, 'Gaseous Air Pollutants and Plant Metabolism', eds. M.J.Koziol and F.R.Whatley, Butterworth, 1984, Chpt.1, p.3.

M.W. Holdgate, 'A Perspective of Environmental Pollution', Cambridge University Press, Cambridge, 1979.

J.Widdows and D.Johnson, 'Biological Effects of Pollutants',ed. B.L.Bayne, K.R.Clarke, and J.S.Gray. Marine Ecology Progress Series Special Issue, Amelinghausen : Inter Research, 1988, 46.

T.W.Clarkson, L. Friberg, G.F. Nordberg, and P.R.Sagar (ed), 'Biological Monitoring of Toxic Metals', Plenum, New York and London, 1988.

L.Friberg, G.F.Nordberg, and V.B. Vouk (ed.s), 'Handbook on the Toxicology of Metals', VII, 2nd Edn., Elsevier, 1986.

World Health Organization, Geneva, Environmental Health Criteria, 'Mercury,' 1976, 1.

H.L.Needleman (ed.), 'Low Level Lead Exposure : The Clinical Implications of Current Research', Raven New York, 1980.

World Health Organization, Geneva, Environmental Health Criteria, 'Photochemical Oxidants', 1979, 9.

D.W.Schindler, T.M.Frost, K.H.Mills, P.S.S.Chang, I.J.Davis, L.Findley, D.F.Malley, J.A.Shearer, M.A.Turner, P.J.Garrison, C.J. Watras, K.Webster, J.M.Gunn, P.L.Brezonik, and W.A.Swenson, in 'Acid Deposition—Its Nature and Impacts', eds. F.T.Last and B. Watling, Proc. Roy. Soc. Edinburgh, 1991, Section B,97,P.193.

World Health Organization, Geneva, Environmental Health Criteria, 'Methylmercury', 1990, 101.

CHAPTER 10

Biogeochemical Cycles

Introduction

Important exchanges of mass and energy occur at the boundaries of the compartments and many processes of great scientific interest and environmental importance occur at these interfaces. A physical example is that of transfer of heat between the ocean surfaces and the atmosphere, which has a major impact upon climate and a great influence upon the general circulation of the atmosphere. A chemically-based example is the oceanic release of dimethyl sulphide to the atmosphere, which may, through its decomposition products, act as a climate regulator.

Pollutants emitted into one environmental compartment will, unless carefully controlled, enter others. Fig 10.1. illustrates the processes affecting a pollutant discharged into the atmosphere. As mixing processes dilute it, it may undergo chemical and physical transformations before depositing in rain or snow (wet deposition) or as dry gas or particles (dry deposition). The deposition processes cause pollution of land, freshwater, or the seas, according to where they occur. Similarly, pollutants discharged into a river will, unless degraded, enter the seas. Solid wastes are often disposed into a landfill. Nowadays these are carefully designed to avoid leaching by rain and dissemination of pollutants into groundwaters, which might subsequently be used for potable supply. In the past, however, instances have come to light where insufficient attention was paid to the potential for groundwater contamination, and serious pollution has arisen as a result.

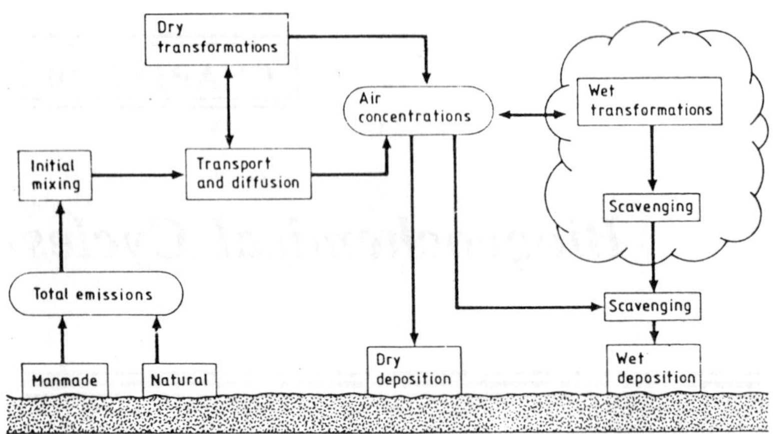

Fig. 10.1 *Schematic diagram of the atmospheric cycle of a pollutant*

Another important consideration regarding pollutant cycling is that of degradability, be it chemical or biological. Chemical elements (other than radioisotopic forms) are, of course, non-degradable and hence once dispersed in the environment will always be there, although they may move between compartments. Thus, lead, for example, after emission from industry or motor vehicles, has a rather short lifetime in the atmosphere, but upon deposition causes pollution of vegetation, soils, and water. On a very long timescale, lead in these compartments will leach out from soils and transfer to the oceans, where it will concentrate in bottom sediments.

Some chemical elements undergo chemical changes during environmental cycling which completely alter their properties. For example, nitrate added to soil as fertilizer can be converted to gaseous nitrous oxide by biological denitrification processes. Nitrous oxide is an unreactive gas with a long atmospheric lifetime which is destroyed only by breakdown in the stratosphere. As will be seen later, nitrogen in the environment may be present in a wide range of valence states, each conferring different properties.

Some chemical compounds are degradable in the environment. For example, methane (an important greenhouse gas) is oxidized via carbon monoxide to carbon dioxide and water. Thus, although the chemical elements are conserved, methane itself is destroyed and were it not continuously replenished, would disappear from the atmosphere. The breakdown of methane is an important source of water vapour in the stratosphere, illustrating another, perhaps less obvious, connection between the cycles of different compounds.

Table 10.1 Size and vertical mixing of various reservoirs

	Mass (kg)	*Mixing time (years)*
Biosphere*	4.2×10^{15}	60
Atmosphere	5.2×10^{18}	<0.2
Hydrosphere	2.4×10^{21}	1600
Crust	2.4×10^{22}	$>3 \times 10^{7}$
Mantle	4.0×10^{24}	$>10^{8}$
Core	1.9×10^{24}	

*Plants, animals, and organic matter are included but coal and sedimentary carbon are not. The mixing time of carbon in living matter is about 50 years.

Source : P. Brimblecombe, 'Air Composition and Chemistry', Cambridge University Press, Cambridge, 1986.

Degradable chemicals which cease to be used will disappear from the environment. PCBs are no longer used industrially to any significant degree, having been replaced by more environmentally acceptable alternatives. Their concentrations in the environment are decreasing, although because of their slow degradability (*i.e.* persistence), it will take many years before their levels decrease below analytical detection limits.

The transfer of an element between different environmental compartments, involving both chemical and biological processes is termed biogeochemical cycling. The biogeochemical cycles of the elements lead and nitrogen will be discussed later in this chapter.

Environmental Reservoirs

To understand pollutant behaviour and biogeochemical cycling on a global scale, it is important to appreciate the size and mixing times of the different reservoirs. These are given in Table 10.1 The mixing times are a very approximate indication of the timescale of vertical mixing of the reservoir. Global mixing can take very much longer as this involves some very slow processes. These mixing times should be treated with considerable caution as they oversimplify a complex system. Thus, for example, a pollutant gas emitted at ground level mixes in the boundary layer (*ca.* 1 km) on a timescale typically of hours. Mixing into the free troposphere (1-10 km) takes days, whilst mixing into the stratosphere (10-50 km) is on the timescale of several years. Thus, no one timescale describes atmospheric vertical mixing and the same applies to other reservoirs. Such concepts are useful, however, when considering the behaviour of trace components. For example, a highly reactive hydrocarbon emitted at ground level will probably be decomposed in the boundary layer. Sulphur dioxide, with an atmospheric lifetime of days, may enter the free troposphere but in unlikely to enter the stratosphere. Methane, with a lifetime of several years, extends through all of the three regions.

It should be noted from Table 10.1 that the atmosphere is a much smaller reservoir in terms of mass than the others. The implication is that a given pollutant mass injected into the atmosphere will represent a much larger proportion of total mass than in other reservoirs. Because of this, and the rather rapid mixing of the atmosphere, global pollution problems have become serious in relation to the atmosphere before doing so in other environmental media. The converse also tends to be true, that once emissions into the atmosphere cease, or diminish, the beneficial impact is seen on a relatively short time-scale.

Residence Time

A very useful concept in the context of pollutant cycling is that of the lifetime of a substance in a given reservoir. We can think in terms of substances having sources, magnitude *S*, and sinks, magnitude *R*. At equilibrium :

$$R = S$$

An analogy is with a bath ; the inflow from a tap (S) is equal to the outflow (R) when the bath is full. An increase in S is balanced by an increase in R. If the total amount of substance in the reservoir (analogy = mass of water in the bath) is A, then the residence time, τ is defined by :

$$\tau = \frac{A\ (\text{kg})}{S\ (kgy^{-1})} \qquad (1)$$

It the removal mechanism is a chemical reaction, its rate may be described as follows :

$$R' = \frac{d[A]}{dt} = k[A]. \qquad (2)$$

(In this case d[*A*]/d*t* describes the rate of loss of A if the source is switched off; obviously with the source on, at equilibrium d[*A*]/d*t* = 0).

The latter part of equation (2) assumes first order decay kinetics, i.e. the rate of decay is equal to the concentration of A, termed $[A\ ]$, multiplied by a rate constant, k. This is often a reasonable approximation.

Taking equation (1) and dividing both numerator and denominator by the volume of the reservoir, allows it to be re-written in terms of concentration. Thus :

$$\tau = \frac{[A\ ]\ (kg\ m^{-3})}{S'\ (kg\ \mathrm{m}^{-3}\ s^{-1})} \tag{3}$$

since $S' = R'$

$$\tau = \frac{[A\ ]}{k\ [A\ ]} = k^{-1} \tag{4}$$

Thus the residence time of a constituent with a first order removal process is equal to the inverse of the first order rate constant for its removal. Taking an example from atmospheric chemistry, the major removal mechanism for many trace gases is reaction with the hydroxyl radical, OH. Considering two substances with very different rate constants for this reaction, methane and nitrogen dioxide :

$$\mathrm{CH_4 + OH \rightarrow CH_3 + H_2\,O} \tag{5}$$

$$\frac{-\mathrm{d}}{\mathrm{d}t}[\mathrm{CH_4}] = \mathrm{k_2\,[CH_4]\,[OH]}\quad \mathrm{k_2} = 8.4 \times 10^{-15}\ \mathrm{cm^3\ molecs^{-1}} \tag{6}$$

$$\mathrm{NO_2 + OH \rightarrow HNO_3}\quad \mathrm{k_2} = 1.1 \times 10^{-11}\ \mathrm{cm^3\ molecs^{-1}} \tag{7}$$

Making the crude assumption of a constant concentration of OH radical (more justifiable for the long-lived methane, for which fluctuations in OH will average out, than for short-lived nitrogen dioxide),

$$\frac{-\mathrm{d}}{\mathrm{d}t}[\mathrm{CH_4}] = k_2\,[\mathrm{CH_4}]\,[\mathrm{OH}]$$

$$= k'_1\,[\mathrm{CH_4}]$$

where $k'_1 = k_2\,[\mathrm{OH}]$

$$= 8.4 \times 10^{-15} \times 1 \times 10^{6}$$

$$= 8.4 \times 10^{-9}\ s^{-1}$$

assuming an OH concentration[6] of 1×10^{6} molec cm^{-3}.

Then from equation (4)

$$\tau = k^{-1}$$

$$\approx (8.4 \times 10^{-9})^{-1} s$$

$$= 3.8 \text{ years for } CH_4$$

By analogy, for nitrogen dioxide, the residence,

$$\tau = 25 \text{ hours}$$

This general approach to atmospheric chemical cycling has proved useful in many instances. For example, measurements of atmospheric concentration, [A], for a globally mixed component may be used to estimate source strength, since

$$S' = R' = \frac{d[A]}{dt} = k_2 [A][OH]$$

and

$$S = S' \cdot V$$

where V is the volume of atmosphere in which the component is mixed. Source strengths estimated in this way, for example for the compound methyl chloroform, CH_3CCl_3, known to destroy stratospheric ozone, may be compared with known industrial emissions to deduce whether natural sources contribute to the atmospheric burden.

THE CARBON CYCLE

The carbon cycle is mainly associated with living matter, although inorganic carbon provides important segments to the complete picture. Atomic number of carbon is 6 and atomic weight 12. About 1 per cent of carbon has an atomic weight of 13, and is often used as a tracer to determine sources. A minute amount occurs as carbon-14, a radioactive isotope used for dating purposes. Fig 10.2. presents the global carbon cycle and Table 10.2 an estimate of the mass content for each part of the cycle.

The cycling of carbon is controlled by storage is reservoirs. The time of such storage may range from millenia for rocks, through decades for the deep ocean layers, to seasons for active biota. Warneck (1988) has suggested the following time periods : geochemical activity involving rocks, 2,400 to 30,000 years ; deep ocean layers, 20 years ; ocean mixed layer, 4-10 years : soil humus, 200 years ; long-term biosphere storage, 75 years ; and short-term biosphere storage, 15 years.

TABLE 10.2 : Estimates of the Mass Content of Carbon in Various Global Reservoirs

Reservoir	Mass Content of Carbon (Pg)
Oceans	
Total dissolved CO_2	37400
Dissolved CO_2 in mixed layer (75 m depth)	670
Living biomass carbon	3
Dissolved organic carbon	1000
Sediments	
Continental and shelf carbonates	270×10^5
Carbonates in ocean	230×10^5
Continental and shelf organic carbon	100×10^5
Organic carbon in ocean	200×10^4
Biosphere	
Terrestrial biomass	650
Soil organic	2000
Oceanic organic	1000
*Atmosphere**	
Present level (350 ppmv)	734
Pre-industrial (290 ppmv)	615

* Virtually all as CO_2.

Source : Warneck (1988).

The Oceans

The major storage of carbon in the oceans occurs in the intermediate and deep water, below the thermocline. The deep layers of the ocean have a very slow mixing period, with carbon remaining in situ for at least 20 years. The mixed layer (mixing depth is generally assumed to be 75 m) provides the main medium of interchange with the atmosphere, where carbon storage is about one and a half times lower. Ninety per cent of the carbon in the ocean is stored as bicarbonate ((HCO_3^-) and about 90 per cent as carbonate (CO_3^-).

The average concentration of inorganic CO_2 in the ocean surface layer is 2.05 mmol m^{-3}, which increase rapidly with depth of about

2.29 mmol m^{-3} at 1 km, and remain fairly constant thereafter. Variations occur with the temperature of the ocean water, as a colder water is about to dissolve more CO_2.

Organic carbon occurs from precipitated skeletons and remains of phytoplankton and other similar arganisms. About 80 per cent of the precipitated material may be redissolved in deep ocean layers. The dissolved organic carbon content of ocean waters is estimated to be 0.7 gm^{-3}. The remaining carbon is particulate, mainly calcium carbonate ($CaCO_3$).

Sediments and Rocks

Carbon constitutes only 0.032 per cent of the earth's crust be mass. Over very long periods of time, the carbon in terrestrial rocks is dissolved by rain or surface water and carried by runoff to be deposited on continental shelf sediments. In the deeper oceans, deposits from organisms build up on the floor over millenia, where carbonate mass more than four times higher than the oceans or in the biosphere remains locked. The exchange of carbon from these locations occurs over thousands of years. About two-thirds of this carbon is inorganic, and the rest organic.

The Biosphere

The exchanges of carbon in the biosphere occur through living plants and animals ; through release of carbon from leaf litter, debris, and other dead biota ; and from soil humus. The mass of carbon in humus is nearly three times higher than in the living biosphere Table 10.2. However, exchange processes are generally inactive and storage may occur for 200 years. Plants use CO_2 as a basic building block for their growth and survival, mainly through photosynthesis.

The Atmosphere

Exchanges of carbon with the atmosphere occur mainly through the biosphere, with the ocean mixed layer an important secondary source. The most important gas is CO_2. Global estimates of the exchange of CO_2 between the atmosphere and biosphere are 113 Pg y^{-1} for assimilation of CO_2 into plants, 55 Pg y^{-1} for respiration, 42 Pg y^{-1} for microbial decay, 5 Pg y^{-1} for herbivore consumption, 10 Pg y^{-1} from soil humus, and 1 Pg y^{-1} from forest fires and agricultural burning (Warneck 1988).

The global carbon cycle is shown in Fig. 10.2. The assimilation and respiration (both marine and biota) associated with CO_2 are much larger than any of the other factors. The other more minor gases in the carbon chain are : carbon monoxide (CO), methane (CH_4), and non-methane hydrocarbons (represented by formaldehyde-HCHO). CO_2 is relatively inert, while the other are active in global atmospheric chemistry processes.

CO_2 is an infrared radiation absorber and has a major impact on the heat balance of the globe . Presentday concentrations are about 350 ppmv. The total mass of carbon in CO_2 in the atmosphere is about 754 Pg. Before the industrial revolution, the mass was about 615 Pg. The increase of about 140 Pg of carbon since about 1860, has been blamed on anthropogenic emissions. Fossil fuel combustion provides more than double the anthropogenic CO_2 compared to all other sources combined. Anthropogenic carbon contributes about 3 per cent to the annual carbon loading of the atmosphere.

Other carbon compounds in the atmosphere are : CO, CH_4 non-methane hydrocarbons, particulate organic carbon (POC), and elemental carbon (EC). About 90 per cent of CO originates from photochemical production of CH_4 with smaller amounts from biomass burning and oxidation of organic gases emitted from vegetation (NMHC). CO is removed from the atmosphere mainly by oxidation to CO_2.

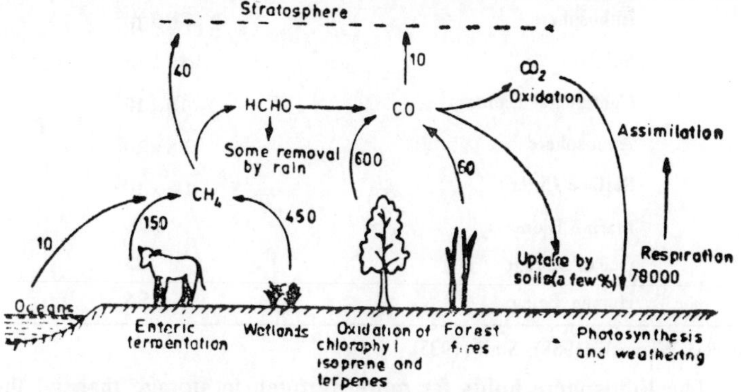

Fig. 10.2 : The global cycle of carbon. (Amounts are in Tg of carbon per year).

CH_4 is released from rice paddies, wetland areas, enteric fermentation from animals, and biomass burning. It has an average concentration of 1.6 ppmv. Sinks include temperate and tropical soils and oxidation to CO. Non-methane hydrocarbons represent a complex set of hydrocarbons with highly varying characteristics, most of which are active chemically and have short lifetimes. Concentrations tend to be only a few ppbv. Sinks are usually associated with photochemical reactions. POCs are a complex mixture of hydrocarbons, alcohols, esters, and organics in particulate form. They usually generate from secondary reactions (gas-to-particle conersion), and are important in cloud and precipitation processes.

Elemental carbon comes almost exclusively from biomass and fossil fuel combustion, is a fine black powder, and is an excellent tracer for long range transport from anthropogenic sources.

THE NITROGEN CYCLE

Nitrogen, and its exchange between the biosphere and the atmosphere, is more essential to ecosystem survival than any other element. Nitrogen, N_2, is an inert gas comprising 78 per cent of the atmosphere. N_2 is crucial to nitrogen use by ecosystems, but several other nitrogen compounds, are also important. Nitrogen and its compounds are essential building blocks for ecosystem and human survival. The global estimates of total nitrogen stored in atmospheric and surface locations are given in Table 10.3.

TABLE 10.3 : Major Areas of Nitrogen Storage in Ecosystems and Estimated Amounts on a Global Scale

Location	Nitrogen Starage (Tg)
Lithosphere	$2-6 \times 10^6$
Soil	85×10^3
Continental Biomass	10×10^3
Atmosphere	3.8×10^3
Surface Litter	1.5×10^3
Marine Biomass	380
Ocean Storage	23
Human Beings	5.5

Source : Warneck (1988); Smil (1985).

The lithosphere holds far more nitrogen in storage than all the other locations combined. Stored in primary, igneous rocks this nitrogen is not accessible to any use by ecosystems. Only a very small fraction is released by weathering and other natural processes. Only small amounts of nitrogen are made available for ecosystem use, and inclusion in the active nitrogen cycle from other regions.

The Biosphere

Fig. 10.3 illustrates the movement of nitrogen in the biosphere. The major active zone of nitrogen use and transfer occurs in the soil and biosphere on the continents. Nitrogen is made available for plant use through the processes of **ixation** which converts mainly inert, inactive N_2 to the active organic from through bacterial processes. This active nitrogen occurs in a form that all plants and micro-organisms can use. Most of the N_2 made available for fixation comes from the atmosphere and from the decay of free-living organisms. Nitrogen is then given to the plants in an assimilatable from through mineralisation to ammonia (NH_3), or through the oxidation of reduced NH_3 to nitrate (NO_3^-) through several chemical steps. This process known as nitrification occurs under aerobic conditions. The cycle in the soil is completed by the return of oxidised nitrogen to an inert condition through **denitrification,** under anaerobic conditions.

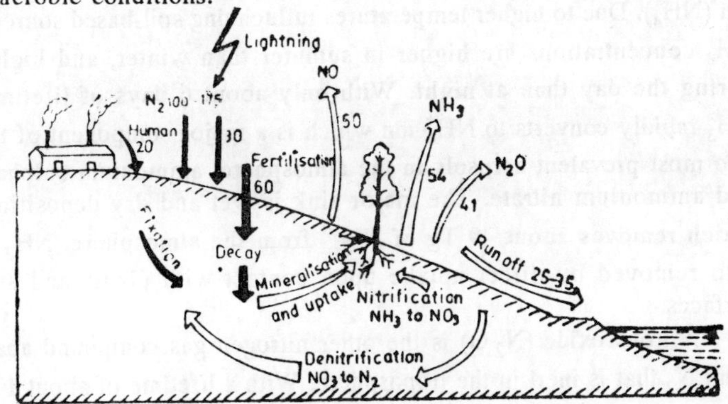

Fig. 10.3 : The cycle of nitrogen in the soil and the resultant release of nitrogen compounds to the atmosphere estimates from Warneck 1988 and Smil 1985.

Aerobic processes also lead to the formation of nitrogen dioxide (NO_2), most of which is released to the atmosphere. Anaerobic processes create nitric oxide (NO), nitrous oxide (N_2O), and N_2. All of these pathways and processes depend on the environmental conditions such as soil acidity (pH < 6 inhibits denitrification), water content, soil type, ad plant type, amount, and temperature. Biochemical activity increases exponentially with increasing temperature.

Surface and Ocean Water

The content of nitrogen in ocean and continental surface waters is much less than that in the biosphere or atmosphere . Over 95 per cent of nitrogen stored in the ocean is in molecular form and is inactive. Only nitrate (about 2.5% of total ocean nitrogen) and organic matter (1.5%) have a small active role. Oceans gain nitrogen through river runoff from continents and wet and dry deposition from the atmosphere. Nitrogen is lost through deposition to bottom sediments, and releases to the atmosphere in areas of biological activity.

Nitrogen in the Atmosphere

As the only gas soluble in water, ammonia (NH_3) is a very important component of the nitrogen cycle because it can directly act as a nutrient for the biosphere. The aerosol component is ammonium ion (NH_4^+). Due to higher temperatures influencing soil-based sources, NH_3 concentrations are higher in summer than winter, and higher during the day than at night. With only about 6 days of lifetime, NH_3 rapidly converts to NH_4^+ ion which is a major component of the two most prevalent aerosols in the atmosphere, ammonium sulphate and ammonium nitrate. The major sink is wet and dry deposition, which removes about 49 Tg of N y^{-1} from the atmosphere. NH_3 is also removed by direct uptake upon contact with plants and soil surfaces.

Nitrous oxide (N_2O) is the other nitrogen gas compound apart from N_2 that is inert in the troposphere. With a lifetime of about 170 years, its major sink is photochemistry in the stratosphere, where it may interfere with the ozone layer. N_2O is also a greenhouse gas. The major source of N_2O is emissions from soil and oceans through microbial processes. Anthropogenic inputs are only about 8 per cent of natural emissions, but are considered important because they are increasing. N_2O emissions increase with higher temperature and moisture, reach a daily maximum around noon, and seasonally the maximum occurs in summer. Emissions can be considerably enhanced on a local scale by irrigation practices. With its long lifetime and major natural sources, N_2O shows very little variation in global distribution, and decreases slightly with height in the troposphere, depending on the availability of photochemical activity.

The nitrogen compounds that are most strongly influenced by anthropogenic emissions are **nitric oxide** (NO) and **nitrogen dioxide** (NO_2). These are often considered in combination as NO_x. Of the two gases, the main primary emission is NO, but this rapidly oxidises to NO_2, leaving the latter dominant in the atmosphere. NO_2 is short lived and oxidises rapidly into NO_3^- aerosol or nitric acid HNO_3. It is crucial to the formation of photochemical smog, peroxyacetyl nitrate (PAN) and other oxidants, and is a major component in the formation of tropospheric and stratospheric ozone.

The major sources of NO_2 are from the burning of fossil fuels or from biomass burning. In North America, emissions from fossil fuels exceed those from natural sources by 3.13 times. The amount of nitrogen released by the burning of fossil fuels depends on the temperature of the burning process and the nitrogen content of the fuel.

Natural sources include soil releases and lightning with a global emission of about 50 Tg y^{-1}, NO_2 contributes about one-third of the nitrogen gained by the atmosphere. About 43 Tg Ny^{-1} are removed, almost entirely through wet and dry deposition, with smaller amounts lost to photochemical reactions to other species.

Atmospheric concentrations of NO_2 is clean ocean air in the troposphere are < 100 pptv. Over the continents, rural air carries about 200-300 pptv, and in air influenced by anthropogenic activities, concentrations of > 10 ppbv, perhaps reaching 500 ppbv in urban air, are common. NO^2 decreases rapidly with height, to a background value of 10

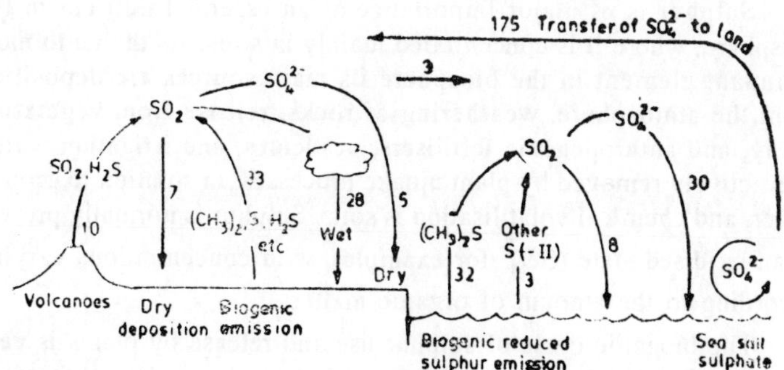

Fig. 10.4. The global sulphur cycle estimated from various sources, in Tg sulphur per year.

pptv in the upper troposphere (Warneck 1988). Higher concentrations of NO^2 occur in the winter, particularly in the mid-latitude areas of the world under urban influence, since temperature inversions are more prevalent and photochemical activity is at a minimum.

THE SULPHUR CYCLE

The main storage of sulphur in the oceans is through dissolved sulphate (SO_4^{2-}), averaging 2.7 g kg^{-1} in ocean water. Oceans are also the major source of dimethyl sulphide (DMS-$(CH_3)_2S$), which is the most abundant volatile sulphur compound in sea water. Produced by algal and bacterial decay, its concentration in sea water is about 100×10^{-9} L^{-1}. The highest concentrations of DMS occur in coastal marshes and wetlands.

The content of sulphur in rocks varies widely according to type, with sedimentary rocks containing about 0.38 per cent, S, but igneous rocks only 0.032 per cent S (Smil 1985). Sulphur is mobilised in the geosphere by slow weathering of rock material caused by precipitation processes. Dissolved in run off, it moves in streams and rivers to the oceans. Sulphur is the second most abundant element in rivers. About 100 Tg of sulphur per year are transported to the oceans, while only 33.42 Tg y^{-1} are weathered from rocks and sediments. The unaccounted portion is likely to be marine sulphur transported landward by the atmosphere and then deposited. The eventual resting place of the sulphur transported to the oceans is mainly in continental shelf sediments, which, eventually on a geologic time scale, would uplift to the surface again, completing the geochemical part of the sulphur cycle.

Sulphur is of major importance as an essential nutrient in the biosphere, where it is concentrated mainly in soils. As the tenth most abundant element in the biosphere its main sources are deposition from the atmosphere, weathering of rocks, release from vegetation decay, and anthropogenic fertilisers, pesticides, and irrigation water. It is actively removed by plant uptake processes, in solution to ground water, and chemical volatilisation is soils. Sulphur is normally present in an oxidised state (SO_4^{2-} for example), with concentrations varying according to the amount of organic matter.

The biogenic cycle of sulphur use and release by plants is very complex. Sulphur in soil may be found in bound or unbound form, as organic or inorganic compounds, organic sulphur is most prevalent. Plants use sulphur to bind complex molecules, such as proteins, enzymes, and antibodies, as part of life processes. Upon the death of the plant, decomposition releases sulphur back into the soil. The action of microorganisms is crucial to the decay process, liberating organic sulphides in two ways. **Aerobic** decay, occurring with free access to oxygen, allows reoxidation of organic sulphides into sulphate, to begin the uptake process all over again. Anaerobic decay, in a soil environment with limited oxygen, binds sulphur in organic form, with eventual release in part to the atmosphere at H_2 S, DMS, and other organic compounds. The approximate amount of sulphur released from global soils is about 7 Tg y^{-1}.

Sulphur in the Atmosphere

The interaction of processes between the earth's surface and the atmosphere results in several sulphur compounds being released into the air. Of these, six compounds most important, are given along with some general information about sources, production, sinks, and average concentration.

The most abundant sulphur species in the atmosphere is carbonyl sulphide (COS). Although small amounts are produced by anthropogenic combustion processes (< 25%), the majority of COS is naturally produced from soil decomposition, marshes and wetlands along oceans coasts, and areas of ocean upwelling that are rich in nutrients.

The only sinks for COS are stratospheric photolysis and slow photochemical reactions in the troposphere. The ocean itself may act as a source and sink, but to what degree is unknown. COS contains about 80 per cent of the total atmospheric S content, but is relatively inert and adds very little to the pollution problems created by sulphur in the atmosphere.

Carbon disulphide (CS_2) has similar sources to COS but on a smaller scale. Far more reactive that COS, its major sink is photochemical reactions, and its lifetime 12 days. The most prominent source of CS_2 is microbial processes in the warm soils of the world. Thus, production of the gas is the greatest in the tropics. The major secondary source is marshes and wetlands along sea coasts. Anthropogenic inputs from automobiles and coal-fired power plants are small.

Dimethyl sulphide is released from oceans in much greater amounts than either COS or CS_2. DMS has a lifetime of only 0.6 days. It is rapidly oxidised to SO_2 or redeposited back into the ocean. In the sulphur cycle, most of the natural gas released is DMS from the oceans. DMS concentrations are higher at night, when sunlight is not present.

Hydrogen sulphide (H_2S), is mainly natural in origin. It has a strong odour, and is created during processes of anaerobic decay is soils, wetlands, salt marshes, and other areas of stagnant water. Maximum concentrations occur over wet tropical forests.

H_2S is a highly active gas (lifetime 4.4 days), and is removed by reaction with the hydroxyl radical (OH) and COS. Highest concentrations occur at night and in the early morning when photochemical activity is at a minimum.

Sulphur compound which is most influenced by anthropogenic emissions is sulphur dioxide (SO_2). About half of global SO_2 originates from the oxidation of H_2S. Concentrations released by the burning of fossil fuels, particularly coal and oil have exceeded natural releases. In some industrialised areas such as eastern North America, over 90 per cent of SO_2 is from anthropogenic sources. A major natural source of SO_2 is volcanic eruptions.

SO_2 has a relatively short lifetime of 2.4 days, and is lost (about 55 per cent) due to photochemical conversion to SO_4^{2-} ion. The rest of the gas lost from the atmosphere (about 45%) is removed by wet and dry deposition. Projections based on fuel use indicate that by the year 2000, in the absence of major international control 150-180 Tg of sulphur per year from SO_2 from anthropogenic sources will be released. This figure is about twice the amount from all natural sources and four times the natural emissions from land based sources.

The largest natural source of sulphur to the atmosphere is sulphate aerosol (SO_4^{2-}) from sea spray. Most falls back into the ocean, but some is carried over the continents. Only 3 Tg y^{-1} of SO_4^{2-} is released directly to the atmosphere from anthropogenic sources. However, much greater amounts are formed through secondary reactions from various sulphur species in the atmosphere.

REFERENCES

V. Smuel *Carbon, Nitrogen, Sulphur*, Plenum, New York 1985. 459 pp.

P. Warneck. *Chemistry of Natural Atmospheres*. vol. 41, International Geophysics Series, Academic Press, New York 1988. 757 pp.

CHAPTER 11

Environmental Monitoring

Objectives of Monitoring

The gathering of information on the existence and concentration of substances in the environment, either naturally occurring or from anthropogenic sources, is achieved by **measurement** of the substance or phenomenon of interest. However, single measurements of this type made in isolation are virtually worthless, since temporal and spatial variations cannot be deduced. Rather, it is necessary to *monitor* the parameter of interest by repeated measurements made over time (and often over space), with sufficient sample density, temporally and spatially, that a realistic assessment of variations and trends may be made.

Monitoring of the environment may be undertaken for a number of reasons and it is important that these be defined before sampling takes place. The generalization that 'monitoring is done is order to gain information about the present levels of harmful or potentially harmful pollutants in discharges to the environment, within the environment itself, or in living creatures (including ourselves) that may be affected by these pollutants' may be expanded as follows :

(a) Monitoring may be carried out to assess pollution effects on man and his environment, and so to identify and possible cause and effect relationship between pollutant concentrations and, for example, health effects of climatic change.

(b) Monitoring may be carried out in order to study and evaluate pollutant interactions and patterns. For example, source apportionment and pollutant pathway studies usually rely on environmental monitoring.

(c) Monitoring may be carried out to assess the need for legislative controls on emissions of pollutants and to ensure compliance with emission standards.

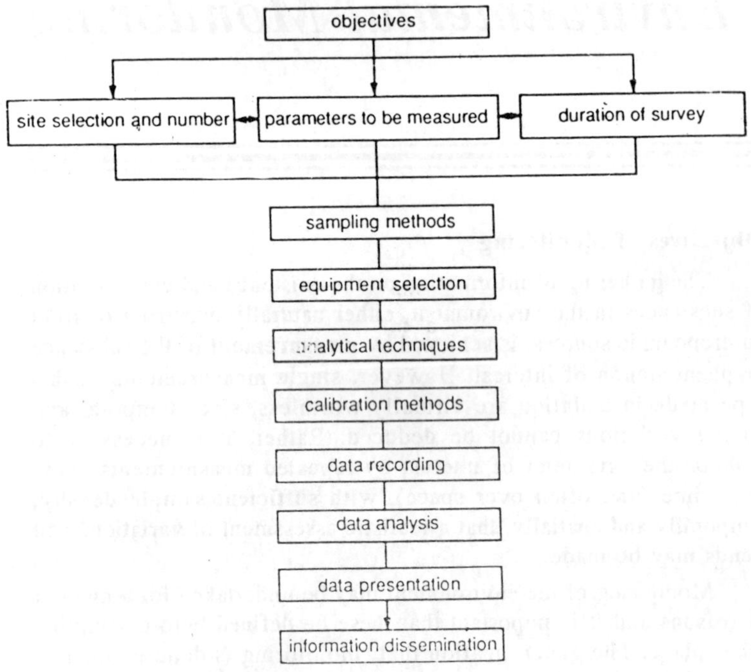

Fig. 11.1. *Steps in the design of a monitoring programme*

An assessment of the effectiveness of pollution legislation and control techniques also depends upon subsequent monitoring.

(d) In areas prone to acute pollution episodes, monitoring may be carried out in order to activate emergency procedures.

(e) Monitoring may be carried out in order to obtain in historical record to environmental quality and so provide a database for future use in, for example, epidemiological studies.

(f) Monitoring may also be necessary to ensure the suitability of water supply for a proposed use (industrial or domestic) or to ensure the suitability of land for a proposed use (for example, for housing).

A basic problem in the design of a monitoring programme is that each of the above reasons for carrying out monitoring demands different answers to a number of questions. For example, the number and location of sampling sites, the duration of the survey, and the time-resolution of sampling will all vary according to the use to which the collected data are to be put. Decisions on what to monitor, when and where to monitor, and how to monitor are often made much easier once the purpose of monitoring is clearly defined. Therefore it is most important that the first step in the design of a monitoring programme should be to set out the objectives of the study. Once this has been done then the programme may be designed by consideration of a number of steps in a systematic way (Fig. 11.1) such that the generated data are suitable for the intended use. It is important also that the data produced by a monitoring programme should be continuously appraised in the light of these objectives. In this way, limitations in the design, organization, or execution of the survey may be identified at an early stage.

The aim of this chapter is to present and discuss the most important and relevant considerations that must be taken into account in the design and organization of a monitoring exercise.

Types of Monitoring

The earth is comprised of three distinct media ; the *atmosphere*, the *hydrosphere*, and *the land*. Pollutants can occur in any or all of the solid, liquid, or gaseous phases. However, the environment is not a simple system and consequently each of the three media may contain pollutants in each of the three phases. Monitoring may therefore be required for a particular pollutant in a specific phase in a particular environmental compartment (*e.g.* sulphur dioxide in air) or it may encompass two or more phases and/or media (*e.g.* dissolved and particulate phase metals in water). Pollutants in the environment originate from a multitude of different types of sources and the identification of these is a necessary prerequisite to the design of a monitoring programme. First, pollutant sources may be classified by their spatial distribution as point sources, line sources, or area sources. Point sources include industrial chimneys, liquid waste discharge pipes, and localized toxic waste dumps on land. Line sources may include highways, airline routes, and run-off from agricultural land, while area emissions may arise from extensive urban or industrial complexes.

Sources may be classified as either statutory or mobile, motor vehicles being the obvious example of the latter. Classification may also be made for air pollutant sources on the basis of the height of discharge, *i.e.* at street level, building level, stack level, or above the atmospheric boundary layer level.

A further important distinction may be made between 'planned', 'fugitive', and 'accidental' emissions to the environment :

(a) Planned emissions arise when (as is invariably the case) it is economically or technically impossible to completely remove all the contaminants in a discharge and hence the process operation allows pollutants to be discharged to the environment at known and controlled rates. Obvious examples of planned emissions include sulphur dioxide from power generation plants and low-level radioactive effluent during nuclear fuel reprocessing.

(b) Fugitive emissions arise when pollutants are released in an unplanned way, normally without first passing through the entire process. They therefore occur at a point sooner in the process that the stack or duct designed for 'planned' emissions. They generally originate from operations which are uneconomic or impractical to control, have poor physical arrangements for effluent control, or are poorly maintained or managed. An example is the escape of heavy metal contaminated dust from a lead works on vehicle tyres, arising from poor dust control and wheel washing arrangements.

(c) Accidental emissions result from plant failure, such as a burst filter bag or faulty valve, or from an accident involving either equipment or operator error (*e.g.* the Chernobyl reactor accident). Accidental emissions can give rise to very high concentrations but they normally occur only infrequently.

Classification of the sources of pollutants in this way allows the distinction of two differing approaches to their monitoring. On the one hand, samples may be taken of the effluent before discharge to, and dispersion in, the environment without consideration of source strengths and rates. Alternatively, samples may be taken from the ambient environment without consideration of source strengths and rates. Obviously neither one of these approaches alone can necessarily provide all the data required to resolve a particular problem and often it is desirable to complement one with the other.

SOURCE MONITORING

Stationary Source Sampling for Gaseous Emissions

A common feature of many industrial processes is that effluent output exhibits cyclical patterns. These may be related to working shift arrangements or be a function of the operations involved, but both require that source testing or monitoring be planned accordingly. Process operations should be reviewed so that discharges during the period of sampling are representative of the plant output in order to ensure that the samples themselves are representative of the effluent, and that the final pollutant analysis will be a representative measure of the entire output.

Two requirements have been specified for valid source monitoring. First, the sample should accurately reflect the true magnitude of the pollutant emission at a specific point in the stack at a specific instant of time. This requirement is met by adequate sampling instrument design. Secondly, enough measurements should be obtained over time and space so that their combined result will accurately represent the entire source emission. This requires consideration of the emissions both in time and in space, across the entire cross-section of the stack.

In a circular flue, sampling at the centroids of equal-area annular segments will ensure that emission variations across the stack cross-section are quantified. In a rectangular flue sample points should be located at the centroids of smaller equal-area rectangles. Generally eight or twelve such sampling points are adequate to compensate for any deficiencies in the location of the sampling site with respect to the length of the stack and to non-ideal flow conditions at the site caused by bends, inlets, or outlets. If it is a particulate pollutant which is being sampled within the stack, it is important that an isokinetic sampling regime is maintained.

Mobile Source Sampling for Gaseous Effluents

Vehicle and aircraft emissions are heavily dependent upon the engine operating mode (*i.e.* idling, accelerating, cruising, or decelerating) and the results obtained by sampling must be considered specific to the type of operating cycle used during the test. Emission tests are usually performed with the vehicle on a dynamometer equipped with inertia fly wheels to represent the vehicle weight and brake loading on a level road.

Source Monitoring for Liquid Effluents

Liquid wastes and effluents often tend, like gaseous effluents, to be inhomogeneous and care is needed in selecting sampling position. Having considered the location of the site in relation to plant operation (*e.g.* should the site be before or after a particular stage of the process of treatment) it is desirable that a region of high turbulence and/or good mixing be chosen. As for gaseous emissions, several samples may have to be taken across the cross-section of a pipe or channel. Sampling from vertical pipes is less liable to be affected by deposition of solids than sampling from horizontal pipes, and a distance of approximately 25 pipe-diameters downstream from the last inflow should ensure that mixing of the two streams is essentially complete. If suitable homogeneous regions for sampling cannot be found, particularly where suspended materials are present, samples may have to be taken from several positions along the effluent stream.

Where the composition of a liquid effluent is known to very with time, grab samples may be collected at set intervals, either manually or by use of an automatic sampler. A alternative approach is to sample at intervals varying with the flow rate so that a more representative composite may be obtained.

Source Monitoring for Solid Effluents

Solid effluents may arise from a number of different processes, including sludge after sewage treatment, ash residue from municipal incinerators, or low-grade gypsum from desulfurization plants attached to coal fired power stations. In general, solid wastes are even less homogeneous than either liquid or gaseous effluents. Therefore, great effort must be made to ensure that samples are representative of the bulk waste. Monitoring of sewage sludge is particularly common due to sludge acting as an efficient sorption material for heavy metals. Typically, 80—100% of the input lead in a sewage treatment plant is incorporated into the sludge, resulting in sludge-lead concentrations of 120—3000 μg^{-1}. Consideration must therfore be given to the concentrations of pollutants in the material before it is used as fertilizer, incinerated, dumped at sea, or used as land-fill. It may also be mentioned that the determination of the metal balance of a sewage treatment works in an interesting monitoring exercise which may be necessary when considering the fate of the treated effluent and solid waste

In most countries guidelines exist to control the disposal of sewage sludges to land, usually based primarily upon the zinc, copper, and nickel content of the sludge. Hence considerable quantities of other metals, including lead, may be added to land over a normal 30 year disposal period. In the UK the disposal of lead-rich sewage sludges to land is controlled where direct ingestion by animals of contaminated grass or soil can occur. Where such ingestion can occur only sludges of lead content < 2000 mg kg^{-1} dry weight should be disposed. Otherwise, where the sludge is to be mixed with the soil the total lead disposed over the normal 30 year period should be limited to 1000 kg ha^{-1}, corresponding to about 450 μg kg^{-1} of soil of 200 mm depth.

Until fairly recently most trace metal analysis of environmental samples was designed to give a measure of the total elemental concentration in the sample as it was felt that this gave an adequate measure of the pollution load for that metal. It is now realized however that total metal concentrations are often not sufficient and that information based upon some form of physico-chemical speciation scheme is required. This may include, for example, solubility of the pollutant in acids of different strengths, the size distribution of particles, and the association with organic compounds. This is because the physical, chemical, and biological responses to a metal input will vary according to its physical and chemical speciation. One disadvantage of this type of analysis is that it is complicated and time-consuming compared with total metal determinations. Thus speciation studies are invariably limited to a few samples where many (tens or even hundreds) would be taken in a total-metal study.

Ambient Air Monitoring

Air pollution problems vary widely from area to area and from pollutant to pollutant. Differences in meteorology, topography, source characteristics, pollutant behaviour, and legal and administrative constraints mean that monitoring programmes will vary in scope, content, and duration, and the types of station chosen will also vary. However ambient monitoring sites may be divided into several categories :

(a) source-orientated sites for monitoring individual or small groups of emitters as part of a local survey (*e.g.* one particular factory) ;

(b) sites in a more extensive survey which may be located in areas of highest expected pollutant concentrations, high population density, or in rural areas to give a complete nationwide coverage ;

(c) base-line stations to obtain background concentrations, usually in remote or rural areas with no anticipated changes in land-use.

Environmental Water Monitoring

Pollutants enter the aquatic environment from the air (by dry deposition or in precipitation occurring either directly onto the water surface or elsewhere within the catchment area), from the land (either in surface run-off or via sub-surface waters) and directly through effluent discharges (either domestic, industrial, or agricultural). The undesirable effects of pollutants in natural water may be due to :

(a) stimulation of water plant growth—eutrophication—which ultimately leads to deoxygenation of the water and major ecological change ;

(b) their direct or indirect toxic effects on aquatic life ;

(c) the loss of amenity and practical value of the water body, particularly as a source of water for public supply.

Apart from the monitoring of sources of pollutants in liquid effluents sampling may be carried out :

(a) in rivers, lakes, estuaries, and the sea in order to obtain an overall indication of water-quality ;

(b) for rainwater, groundwater, and run-off water (particularly in the urban environment) to assess the influence of pollutant sources

(c) at points where water is taken for supply, to check its suitability for a particular use ;

(d) using sediments and biological samples in order to assess the accumulation of pollutants and as indicators of pollution.

Apart from the measurement of chemical and physical parameters the quantitative or qualitative assessment of aquatic flora and fauna is often used to give an indication of the presence or absence of pollution, and well recognized relationships are known to exist between the abundance and diversity of species and the degree of pollution. This is often used to assess the cleanliness of natural fresh waters and is known as biological monitoring.

Location of sampling sites

There are two main causes of heterogeneous distribution of quality in a water body. These are :

(a) if the system is composed of two or more waters which are not fully mixed (such as in themally stratified lakes or just below an effluent discharge in a river) and ;

(b) if the pollutant distributes non-uniformly in a homogeneous water body (for example oil which tends to float, and suspended solids which tends to settle out of the water). Also chemical and/or biological reactions may occur non-uniformly in different parts of the system, so changing pollutant concentrations non-uniformly. When the degree of mixing is unknown it is advisable to conduct a preliminary survey before deciding on sampling locations. Rapidly obtainable measurements of water temperature, pH, dissolved oxygen, or electrical conductance may be used in this respect.

Sampling locations should generally be at points as representative of the bulk of the water body as possible, *e.g.* away from river or lake banks or the walls of channels or pipes, but often it will be desirable (and necessary) to take samples from several locations in order to obtain the required information.

When sampling from rivers and streams downstream of effluent discharges longitudinal, transverse, and vertical sampling arrays may be necessary to ensure that truly representative data are obtained. Studies of some pollutants require sampling at considerable distances downstream of effluent inputs, *e.g.* in investigating the sag in dissolved oxygen content. When a temporally-varying effluent discharge is under study it may be desirable to sample as close to the point of discharge as mixing allows in order to monitor short-term variations in concentration. However, if long-term average water quality is of interest then sampling should be carried out further downstream where longitudinal dispersion and mixing will have smoothed out the short-term variations.

Sampling in estuaries presents special problems as great spatial and temporal variability may be exhibited. The appropriate locations for sampling will vary from estuary to estuary and will depend on the parameters of interest, but a minimum of 50 samples per survey might be appropriate. If one considers a compound which has an input at one end only of an estuary and which is not removed from or added to solution during its lifetime in the estuary then the concentration of that compound in the estuary will be solely dependent upon the dilution ratio between the river water and the seawater. Thus, for example, the concentration of chloride ion or salinity is dependent only upon the mixing of the fresh and the saline water bodies. This concept, 'conservative' behaviour, is an important one which must be taken into account when monitoring estuarine concentrations. If a graph is drawn of the concentrations of the element of interest in an estuary against salinity then the data points will fall on a straight line (the theoretical dilution line) if physical mixing is the only process controlling the concentration of the element in the water. However, if the element of interest is added to or removed from the solution during mixing then the data will not plot on a straight line. In the case of lakes and reservoirs vertical stratification of pollutants may be very pronounced due to a reduction in dissolved oxygen from the surface downwards. A minimum of three samples is then considered necessary, at 1 m below the surface, 1 m above the bottom, and at an intermediate point.

When water is abstracted from a river, lake, reservoir, or from an aquifer samples should be regularly taken at the point of abstraction and at the point where the water enters the distribution system. Several excellent handbooks with full descriptions of water sampling and analytical methods are available.

The determination of concentrations of trace metals in natural waters is a fundamental stage in the calculation of their budgets or cycles, but is subject to the same problems of sample contamination as occur for atmospheric samples from remote areas. All stages of the analysis, from sample collection, storage, and filtration to actual laboratory manipulation require care to prevent contamination occurring. Indeed for many years measured levels of many trace elements in seawater were purely an artefact of contamination during sampling and analysis.

Sediment, Soil, and Biological Monitoring

Soils and sediments may become polluted by a number of routes, including the disposal of industrial and domestic solid wastes, wet and dry deposition from the atmosphere, and infiltration by contaminated waters.

The main pollution hazards on land have been identified as follows :

(a) Harmful substances may get into the soil or plants and so into the food supply.

(b) Substances may wash from the land and so pollute water supplies.

(c) Contaminants may be resuspended and subsequently inhaled.

(d) Substances polluting the land may make it potentially dangerous or unsuitable for future use (*e.g.* for housing or agriculture).

(e) Ecological systems may be damaged, with consequent loss to conservation and amenity.

Some potentially harmful substances, such as mercury or lead, are naturally present in soils but at concentrations which are not normally deleterious. Some activities however can cause elevated levels of these compounds. For example, mining may cause soils to be contaminated by metals, and the dumping of solid wastes on land will invariably introduce a wide variety of pollutants to the soil. On the other hand there are compounds which do not occur naturally, and their presence in soils and sediments is due entirely to man's activities. These substances include pesticides (particularly the organo-chlorine compounds such as DDT, toxaphene, aldrin, dieldrin) and artificial radionuclides (*e.g.* ^{137}Cs, ^{106}Ru).

As with the other types of media discussed above it is important that the background levels of pollutants be established in soils, sediments, and vegetation. Stream sediments are considered to represent a close approximation to a composite sample of the weathering and erosion products of rock and soil upstream of the sampling point and in the absence of pollution provide information on the regional distribution of the elements.

A second type of monitoring programme is required to establish actual levels of contamination in land or sediments known or believed to be affected by pollutants. In this case much more specific and localized monitoring may be required in order to quantify the degree of contamination. The contamination of sites often arises from their previous uses, particularly as coal-gas manufacturing plants, sewage works, smelters, waste disposal sites, chemical plants, and scrapyards. A typical contaminated site may be found to contain variable concentrations of toxic elements and organic compounds, phenols, coal-tars, and oils, combustible material from undecomposed refuse, acidic or alkaline waste sludges, and sometimes methane accumulations. In addition, contaminated land is often formed by waste tipping and so may be poorly compacted and very unhomogeneous.

Some sites may contain underground pipework and structures from their previous uses and so present formidable sampling problems. Site investigation of this type is very expensive, involving bore holes or trial pits, and, in cases where a detailed site history is unavailable, determination of a large number of pollutants. However it is most important to be sure that the investigation is sufficiently rigorous as remedial measures are extremely costly and must be based upon adequate data. Common problems which have been identified are :

—an inadequate number of samples ;

—an inadequate range of determinands ;

—bulking of samples when individual samples from specific locations are preferable ;

—inappropriate analytical methods ;

—inadequate referencing of sample locations ;

—inadequate descriptions of samples ;

—inadequate descriptions of trial pit strata ;

—an ignorance of the nature of the required information.

Monitoring should be carried out following the application of sewage sludge or waste waters to agricultural land. Samples of surface water, groundwater, site soil, vegetation, and the sludge applied would normally be tested for faecal coliform, nutrients, heavy metals, and pH. The results from the monitoring exercise may be compared to predicted levels derived from the application rates of sludges to land, soil type, nitrogen, phosphorus, and heavy-metal contents of the waste, and the nutrient-uptake characteristics of the cover crop.

When monitoring background levels and more specific pollution on land or in the sediments of a water body, measurements will often be made of levels in the plants or organisms that the soil or sediments support. In many cases flora or fauna provide excellent indicators of the degree of pollution as they may act as bioconcentrators (for example of heavy metals from suspended material in shellfish). Furthermore it is obviously important to monitor pollution levels in food and through the food chain. The simultaneous measurements of pollutant levels in soils and plants as well as in water, sediments, and aquatic biota are therefore often carried out. However the relationships between levels in these various media are often not simple and sampling and analysis of one of these is no substitute for a comprehensive monitoring programme.

Monitoring of concentrations of trace metals in crop plants grown on sewage sludge-amended soils indicates that the levels found will vary with crop species and properties of the soil substrate on which they have been grown. More recent studies have concentrated on the physico-chemical speciation of the metals in the applied material and the receiving soil and have demonstrated the importance of organic complexes in reducing free metal activity.

AIR SAMPLING METHODS

Sampling systems for airborne pollutants usually consist of four component parts, the intake component, the collection or sensing component, the flow

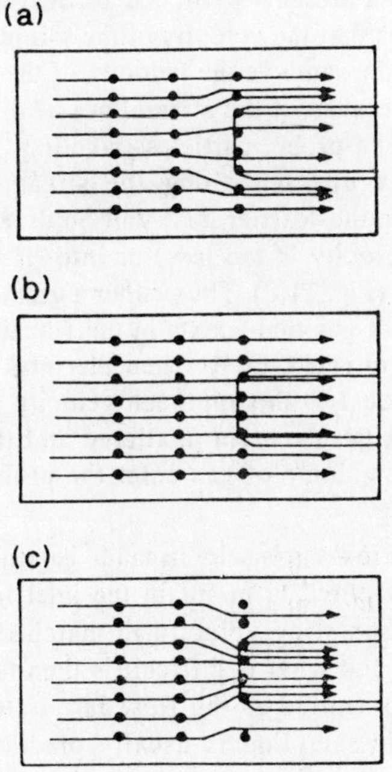

Fig. 11.2. *Schematic diagram of (a) over-sampling of suspended particles ; (b) isokinetic sampling ; and (c) under-sampling of suspended particles*

measuring component, and the air moving device. All these must be constructed of materials which are chemically and physically inert to the sampled air (other than the collector or sensor itself).

Intake design

The nature of the intake is determined by the type and objective of the sampling technique, and may vary from a vertical opening for the passive collection of dustfall in a deposit gauge to a thin-walled probe used for source sampling of aerosols. Common problems which may require consideration are the non-reproducible collection of the sample portion from the air mass due to poor inlet design, adhesion of aerosols to tube walls, loss or change of analyte by chemical reaction with inlet materials, adsorption of gaseous components on inlet materials, and condensation of volatile components within the transfer lines.

The sampling of aerosols presents particular difficulties in inlet design. A basic requirement is that the velocity of the sample entering the system intake should be the same as the velocity of the gas being sampled. This is necessary because, if the streamlines of the sampled gas are disturbed by the intake probe, particles travelling in the gas flow and possessing inertia directed along the streamlines will continue into the probe while the 'carrier gas' will be diverted away from (if the probe intake velocity is too low) or into (if the intake velocity is too high) the inlet (Fig. 11.2). Thus either a greater number of particles per unit volume of gas than exists in the actual gas flow, or a fewer number will be collected. Only when the intake velocity at the face of the probe is equal to the approach velocity of the gas stream will the streamline pattern remain unaltered and the correct number of particles per unit volume of gas enter the probe. This is known as isokinetic sampling.

Sampling of ambient air masses is seldom made isokinetically as sophisticated equipment is required to maintain the inlet facing into the wind and to adjust the sampling velocity to match changes in wind speed. This is feasible, but what is difficult is then to interpret the analysis of the collected sample as the flow rate is temporally variable. However isokinetic sampling is usually practicable, and indeed necessary, when sampling flue gases.

Sample Collection

The methods most commonly used for the collection of atmospheric particulate samples are :

—filtration ;
—impingement : wet or dry impingers, cascade impactors ;
—sedimentation : by gravity in stagnant air, thermal precipitators ;
—centrifugal force, cyclones ;

and for gaseous samples are

—adsorption ;
—absorption ;
—condensation ;
—grab sampling.

Filtration

This is by far the most common technique. The type of filter medium chosen will depend upon a number of factors. These include the collection efficiency for a given particle size, pressure drop and flow characteristics of the filter type, background concentrations of trace constituents within the filter medium, and the chemical and physical suitability of the filter with regard to the sampling environment.

Impingement

Impingers consist of a small jet through which air stream is forced, so increasing the velocity and momentum of suspended particles, followed by an obstructing surface on which the particles will tend to collect. Wet impingers operate with the jet and collection surface under liquid and require high flow rates for optimum collection efficiency.

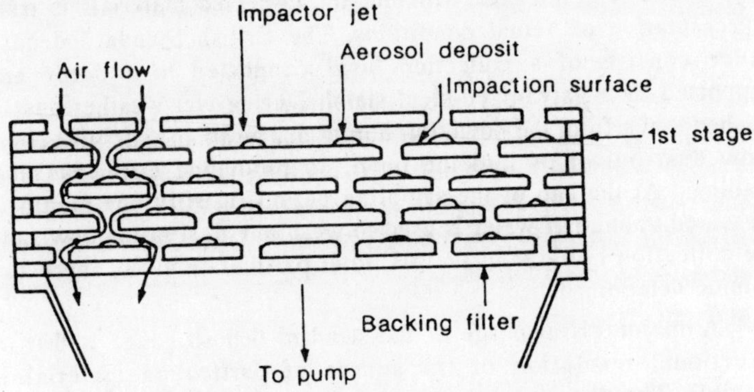

Fig. 11.3. *Schematic representation of a cascade impactor*

Cascade impactors use the aerodynamic impaction properties of particles to separate the sample into different size fractions by use of sequential jets and collection surfaces. Increasing jet velocity at each stage fractionates the sample. Fig. 11.3 shows the principle of a commonly used cascade impactor. This consists of up to seven stages backed by a membrane filter, each stage containing accurately drilled holes which align over a solid portion of the adjacent plates. The holes in each successive stage are smaller than those in the preceding plate, and since air is drawn through the instrument at a constant flow rate the effective velocity at each stage increases. The largest particles are impacted on the first stage and the smallest are collected on the back-up filter. The range of particle diameters collected on each stage may be determined by laboratory calibration or by theoretical calculations. However impaction sampling at normal flow rates ($0.01–0.04\,m^3\,min^{-1}$ or $0.6–1.1\,m^3\,min^{-1}$ for Hi-Vol cascade impactors) and atmospheric pressure is only efficient for particles with aerodynamic diameters >0.3 μm. Also the collection efficiency of each stage will vary according to particle type, some being very 'sticky', others liable to bounce off. Other problems of cascade impactor sampling include wall losses and the aggregation of particles and the mechanical breaking of agglomerates which result in inaccurate size distribution measurements.

Sedimentation

The collection of particulate material by allowing it to deposit into a collection vessel is the simplest of all air pollution measurement techniques. However the presence of the bowl or cylinder in the path of the falling particles will change their flow pattern and it is not clear whether the collected materials is truly representative of actual conditions. The British Standard deposit gauge consists of a collection bowl connected to a bottle and supported by a galvanized steel stand. During wet weather dust is washed down from the bowl, but during dry weather high winds may blow dust out of or into the bowl, so producing erroneous dust loadings. At the end of the sampling period (usually one month) a measured volume of water is used to wash any dust in the bowl into the collection bottle, and the pH, total particulate mass, and water volume determined.

A major disadvantage of the standard deposit gauge is that no directional resolution of the source of particulate material is possible. The directional deposit gauge consists of four cylinders mounted on a common post, with open slots facing the four quadrants of the compass (Fig. 11.4). Each cylinder has a removable collection bottle at its base. After collection a suspension of the dust

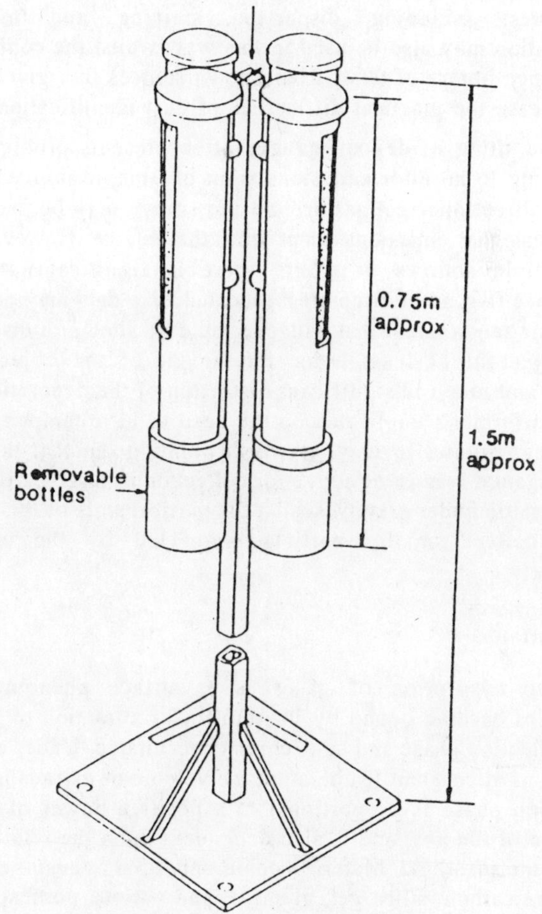

Fig. 11.4. *Directional deposition gauge*

may be placed in a glass cell and a measure of the dust loading made by the amount of obscuration of a beam of light passing through the cell. Density fractionation of the collected materials is also possible by using a mixture of di-iodomethane and acetone which allows a density gradient of 0.8–3.3 g cm^{-3} to be achieved. By comparison of the density gradient fractionation of material collected in a directional deposit gauge with material collected from likely sources (*e.g.* pulverized fly ash storage heaps, slag heaps, cement works, *etc.*) simple source apportionment studies may be attempted. Other techniques, including dispersion staining and microscopic examination may also be used in this way, whilst the compilation of a reference library of dusts from known sources in a given area will greatly ease the practical difficulties of dust identification.

The siting of deposit gauges often presents problems. When attempting to monitor emissions from one major source the use of several directional gauges around the source may be successful in confirming that emissions occur from that source. However in areas of multiple sources or where there is significant atmospheric turbulence (*e.g.* in built-up areas) inconclusive data may be obtained. The basic requirement that gauges should be sited at a distance from any object of at least twice the height of the object is often insufficient to avoid significant distortion of the deposition pattern. Siting is further complicated by the need to have tamper-proof sites at or near ground level. It is worth pointing out that the standard deposit gauge is most effective for collection of large particles which readily settle under gravity, whilst the vertical slots of the directional gauge collect smaller particles impacted by the wind more effectively.

Adsorption

The adsorption of gases is a surface phenomenon. Gas molecules become bound by intermolecular attraction to the surface of a collection phase and so become concentrated. Under equilibrium conditions at constant temperature the volume of gas adsorbed on the collection phase is proportional to a positive power of the partial pressure of the gas, and is also dependent upon the relative surface area of the adsorbent. Materials commonly used as adsorbents include activated carbon, silica gel, alumina, and various porous polymers.

When selecting a suitable adsorbent the relative affinity for polar or non-polar compounds must be considered. For example, activated carbon is non-polar and therefore will absorb non-polar organic gases, but exclude polar compounds such as water vapour. The wide range of gas chromatographic supports available vary in their degree of polarity and so allow selection of the appropriate type.

The adsorbent used must not react chemically with the collected sample unless chemisorption is used intentionally. Also the analytes must not react with other constitutents of the sampled air, either during collection or storage. It has been found for example that tetraalkyl lead may decompose on Porapak Q by reaction with atmospheric ozone. This may be prevented by the use of a selective prefilter which removes the oxidant from the air stream but allows the analytes to pass through.

It is important to determine the retention volume of the adsorbent (*i.e.* the volume of air which may be passed without breakthrough of the analyte) with respect to the species being collected. This should be high enough to allow sufficient of the analytes to be collected for analysis. The desorption properties of the material are also important to ensure quantitative recovery of the sample, preferably with regeneration of the adsorbent for subsequent use. Activated carbon is a very efficient adsorber, so making quantitative desorption difficult. Steam stripping may result in hydrolysis reactions with the analytes and vacuum distillation and solvent extraction are not without their problems. Compounds on support bonded porous polymers may conveniently be thermally desorbed by flushing with an inert carrier gas. In the case of tetraalkyl lead on Porapak Q a two-stage thermal desorption system utilizing an intermediate cryogenic trap cooled with liquid N_2 and then flash-heated to 90°C allows quantitative recovery as well as direct injection of the sample in a very small volume of gas onto the gas chromatograph column.

Since adsorption is temperature-dependent, collection efficiency and an increase in retention volume may be achieved by cooling the adsorbent. However, problems with blockages by ice may then occur. With the increase in sophistication of detection systems in recent years, particularly by the interfacing of chromatographic separation techniques with mass detectors and spectrophotometers, the use of adsorbents as preconcentrators is also increasing. However, care must always be exercised to avoid non-quantitative collection, break-through effect due to exceeding the retention volume of the system, decomposition, and non-quantitative recovery of the sample.

Absorption

Gases may be collected by being dissolved in a liquid collection phase or by chemical reaction with the absorbent. The simple Dreschel bottle may be used or may be modified by the inclusion of a fritted diffuser to create small bubbles and so enhance the collection efficiency.

An example of a simple absorption technique which allows an estimate of NO_2 concentration to be made with relatively little capital outlay is the use of triethanolamine diffusion tubes. An acrylic tube 7.5 cm long × 1 cm i.d. is fitted with a fixed cap at one end and removable cap at the other. A fine wire mesh coated in triethanolamine is placed in the closed end of the tube and adsorbs NO_2 as it diffuses from the open end. The NO_2 is determined spectrophotometrically at the end of the sampling period. These passive samplers are very cheap to construct and analyse and have been used as an effective primary survey technique before embarking upon a more expensive monitoring exercise based upon the standard chemical method or chemiluminescent techniques. In this survey highest NO_2 concentrations were identified at main road junctions.

Condensation

By cooling an air stream to temperatures below the boiling point of the substance of interest it is possible to condense gases from the air and so concentrate them. However, a limitation of the method is that water vapour present in the air will also freeze and so progressively block the trap. This may be overcome by using a first trap of large volume designed to collect water and a second trap at a sufficiently low temperature to collect the analytes. Coolants of temperatures —183°C or lower (*e.g.* liquid N_2) should not be used for this purpose as they will condense atmospheric oxygen and result in a serious combustion hazard.

Grab sampling

Rather than utilizing a concentration technique in the field, samples may be collected in an impermeable container and returned to the laboratory for analysis. Grab samples of this type have been collected in FEP-Teflon bags for hydrocarbon determination by GC for example. In this technique the deflated bag is contained within a rigid box which is slowly evacuated. Air is thus drawn into the flexible bag which may be sealed when inflated. Simples can then be drawn at a later stage from the bag by hypodermic gas-tight syringe.

WATER SAMPLING METHODS

For many applications no special water sampling system is required as an appropriate sample container immersed in the water may be adequate. The main requirement is that a portion of the material under investigation small enough in volume to be transported and handled conveniently but still accurately representing the bulk material should be collected. Typically a 0.5–2 dm^3 volume is sufficient. When samples are required from depth, two types of collection vessel may be used. The first consists of a cylinder with hinged lids at both ends. The container is lowered into the water with both lids opened and at the desired depth a messenger weight sent down the wire which closes them. This type is not suitable for trace metal work as contaminated surface waters may result in contamination of the vessel and the messenger may scour metallic particles from the wire. The second type consists of a sealed container filled with air which is lowered to the required depth. A messenger is again sent down to open, and another to close, the lid. These devices are discussed in further detail elsewhere. Alternatively, pressure sensors may activate the lid.

Automatic sequential samplers are available which will collect a given volume of water into an array of bottles. They have been used, for example, in collecting stormwater run-off from roads and the sampling sequence may be triggered when the flow in a flume reaches a certain height.

A recent innovation in the field of water monitoring is the use of adsorption or filtration media to concentrate the species of interest *in situ*. Using a completely self-contained sealed unit of inert material housing a peristaltic pump and power supply with only the adsorption tube inlet and outlet open to the water, contamination of the sample can be completely avoided. Very large volumes of water may be processed. Future development of this technique employing micro-processor controlled pumps and arrays of collection tubes in moored units is anticipated.

One aspect of water sampling that has attracted a great deal of research is the development of efficient methods of sampling the water surface microlayer, the top 100 μm or so. Early methods centred on the collection of lipids which were known to affect the transfer rates of gases across the air—water interface, and examples of these were the use of a stainless-steel mesh (efficient only for molecules of chain-lenght of 16 carbon atoms or more) and the Harvey Skimmer which picks up a continuous film of water 60—100 μm thick which is then scraped off the rotating drum by a blade. However both these devices collect microscopic surface organisms, known as neustron, as well as the abiotic molecules and lipids which may be leached from the neustron during the subsequent extraction procedure.

A recent development is the use of a germanium slide which will pick up a coherent layer of any surfactants present on the water surface when it is withdrawn perpendicularly. The use of infrared spectrophotometry may then be used to identify the functional groups present, germanium being transparent to infrared radiation. Another method utilizes a large container with an outlet at the bottom and equipped with a mechanical stirrer. As water drains out of the container 'floatables' adhere to the vessel walls and may then be rinsed off with solvent.

When considering sampling methods for use on inland waters or in coastal waters and estuaries the sophisticated techniques developed for use at sea may be found to be impracticable due to their need for heavy lifting gear on the sampling vessel. Monitoring work must often be undertaken on such waters using small boats without such equipment. One ingenious methods of collecting water samples at different depths using very limited resources on a small boat is to lower a weighted plastic tube to the desired depth and to use a small peristaltic pump to draw water up and into a collection bottle. In this way completely uncontaminated samples may easily be obtained from depths of 30 m or more. Whether the particulate fraction of material present in the water can be quantitatively collected in this manner is not clear.

Whichever method of collecting samples is used care must be taken to ensure that neither the sample storage containers nor any collecting vessels used contaminate or alter the sample. This may occur by :

(a) Leaching of contaminants from the surface of imperfectly cleaned containers ;

(b) Leaching of organic substances from plastics or silica and sodium or other metals from glass ;

(c) Adsorption of trace metals onto glass surfaces or organics onto plastic surfaces. In the case of metals this may be avoided by prior acidification of the container, but this may in turn exacerbate problem (a) ;

(d) Reaction of the sample with the container material, *e.g.* fluoride may react with glass ;

(e) change in equilibrium between pollutants in particulate and solution phases.

If a solvent extraction technique is used to concentrate the analytes prior to analysis, care must be taken to ensure that the reagents and containers used are themselves sufficiently clean. Some commonly used materials and techniques have been shown to cause severely elevated metal levels in water.

Different determinands require different methods of preservation in order to prevent significant changes between the time of sampling and of analysis. Generally, acidification to pH 2 and refrigeration to 4 °C will be adequate although complete stability of every constituent can never be guaranteed and, at best, chemical, physical, and biological processes affecting the sample can only be slowed down. Samples may be filtered directly after collection in the field to separate the particulate and solution phases. The solution phase may then be acidified to prevent adsorption of the pollutant to the container wall.

SOIL AND SEDIMENT SAMPLING METHODS

Soils and sediments are typically very inhomogeneous media and large lateral and vertical variations in texture, bulk composition, water content, and pollutant content may be expected. For this reason large numbers of samples may be required to characterize a relatively small area. Although surface scrapings may be taken it is often necessary to obtain cores so that vertical profiles of the determinands may be obtained or cumulative deposition estimated. Plastic or chromium plated steel tubing of 2.5 cm internal diameter is often suitable, and if the samples are sealed into the tubes and air excluded they may be satisfactorily stored at low temperatures until required. Otherwise they may be extruded in the field and stored in plastic bags. Various core sampling devices are available for obtaining cores of bottom sediments from lakes, *etc.* (*e.g.* the Jenkin corer).

Grab samples of soils are easily obtained manually and stored in pre-cleaned plastic bags. Sometimes composite samples formed by the bulking together of a number of individual samples may be sufficient, but generally analyses of individual samples is to be preferred. In the case of sediments, grab samplers are available for operation at considerable depths, examples being the Ponar, Orange-peel, and Peterson grabs. Alternatively a dredge may be used to obtain a composite sample along a strip of the sediment surface.

Wet soils and sediments which are to be analysed while still wet should not be collected or stored in bags, but in rigid containers. The vessel should be filled as completely as possible leaving no airspace at the top and a bung inserted so as to displace excess water without admitting air.

Some determinands in soils and sediments are liable to change during storage and require the use of preservation techniques. For example nitrate in soil can be extracted into potassium chloride solution and preserved with toluene. Usually, however, air-dried soils and sediments may be disaggregated, sub-sampled by coning and quartering and stored in suitable containers, but as always sample contamination must be avoided at each stage.

There are important effects associated with grain size which should be considered in the analysis of soils or sediments. First, many pollutants are associated with particle surfaces and therefore occur in highest concentrations in the smaller grain sized material. Secondly sub-sampling from a bulk sample may be very difficult due to size segregation effects and it may be necessary to grind the sample to a very fine powder to ensure homogeneity prior to division of the sample.